Altitude distribution of vascular plants in mountains of East and Northeast Greenland

FRITZ HANS SCHWARZENBACH

MEDDELELSER OM GRØNLAND, BIOSCIENCE 50

FRITZ HANS SCHWARZENBACH: *Altitude distribution of vascular plants in mountains of East and Northeast Greenland*. Meddelelser om Grønland, Bioscience 50. Copenhagen, the Danish Polar Center, 2000.

Printed by Special-Trykkeriet Viborg a-s
Layout Charlotte Munch and Special-Trykkeriet Viborg a-s
Cover photo Fritz Hans Schwarzenbach

Publishing Editor: Kirsten Caning

Scientific Editor: Dr. Gert Steen Mogensen, University of Copenhagen, Botanical Museum, Gothersgade 130, DK-1123 Copenhagen K. Phone + 45 3532 2200, fax + 45 3532 2210, e-mail gertsm@bot.ku.dk.

About the monographic series Meddelelser om Grønland
Meddelelser om Grønland, which is Danish for *Monographs on Greenland*, has published scientific results from all fields of research in Greenland since 1879. Since 1979 each publication is assigned to one of the three subseries:

- Man & Society
- Geoscience
- Bioscience

Bioscience invites papers that contribute significantly to studies of flora and fauna in Greenland and of ecological problems pertaining to all Greenland environments. Papers primarily concerned with other areas in the Arctic or Atlantic region may be accepted, if the work actually covers Greenland or is of direct importance to continued research in Greenland. Papers dealing with environmental problems and other borderline studies may be referred to any of the series *Geoscience, Bioscience or Man & Society* according to emphasis and editorial policy.

For more information and a list publications, please visit the web site of the Danish Polar Center *www.dpc.dk*

All correspondence concerning this book or the series *Meddelelser om Grønland* (including orders) should be sent to:

Danish Polar Center
Strandgade 100 H
DK-1401 Copenhagen
Denmark
tel +45 3288 0100
fax +45 3288 0101
email dpc@dpc.dk

Accepted December 1999
ISSN 0106-1054
ISBN 87-90369-38-6

Contents

Abstract

An analysis of the altitude distribution of vascular plants is reported for five mountainous areas in East and Northeast Greenland between 71°N and 80°40′ N (field work 1948-1952, 1954, 1956, 1991, 1994-1995, 1998). The analysis rests on a database of 9626 records. There are 176 species documented with 3457 herbarium specimens.

The altitude distribution of each species is shown in bands of 300 m, range 0-1800 m. Unusual altitudinal patterns are suggested to be explained by edaphic, climatic or genetic factors, sometimes also by the history of the vegetation.

A system of "altitude distribution types (ADT)" combines information about the highest sites with the distribution from S to N, and preferred habitats of each species. Biometric estimates of the altitude limit of each species is presented. Limits are highest in the Stauning Alper (72°N, 1720 m), lowest in Kronprins Christian Land (80°N, 900 m), decreasing c. 300 m from the western Stauning Alper to Kong Oscar Fjord.

Average altitude limits of 42 common species are compared with altitude distribution in six other areas in West Greenland, East Greenland and North Greenland. Exposure influences the altitude distribution considerably, depending on the degree of regional climate, continentality and local latitude. The diversity of the vegetation as estimated by Shannon-Index, Simpson-Index, Q-Statistics for the species/presence-distribution is similar in all areas studied.

Keywords: Arctic, Greenland, mountain vegetation, vascular plants, altitude distribution, vertical distribution, altitude limits, climate, ecology, diversity, diversity index, "Vegetationsgrenze".

Fritz Hans Schwarzenbach, Kistlerweg 9, CH-3006 Bern

Zusammenfassung

Präsentiert wird eine Analyse über die Höhenverbreitung der Gefässpflanzen in fünf Berggebieten Ost- und Nordost-Grönlands zwischen 71°30′-80°40′N (Feldarbeit 1948-1952, 1954, 1956, 1991, 1994-1995, 1998). Die Datenbank umfasst 9626 Beobachtungen. Nachgewiesen sind 176 Arten (3457 Herbarbelege).

Die Höhenverbreitung der einzelnen Arten wird nach Höhenstufen von je 300 m im Bereich von 0-1800 m tabelliert. Auffällige Höhenprofile einzelner Arten werden auf klimatische, edaphische oder genetische Faktoren, gelegentlich auch auf die Vegetationsgeschichte zurückgeführt.

Ein Klassifikationssystem „altitude distribution types (ADT)" verknüpft Daten über die Höhen- und Nord/Süd-Verbreitung mit Angaben über die bevorzugten Habitate der einzelnen Arten. Für jede Art werden Schätzungen der vertikalen Verbreitungsgrenzen angegeben.

Mit biometrischen Methoden wird die Vegetationsgrenze bestimmt.. Sie liegt am höchsten in den Stauning Alper (72°N, 1720 m), am niedrigsten in Kronprins Christian Land (80°N, 900 m). Sie sinkt von den westlichen Stauning Alper zur Küste des Kong Oscar Fjordes (72°N) um ca. 300 m ab.

Anhand der Durchschnittswerte der beobachteten Höhengrenzen für 42 häufige Arten werden die Höhenverteilungen der Gefässpflanzen in sechs weiteren Gebiete West-, Ost- und Nordgrönlands verglichen. Die Höhenverteilung hängt von der Exposition der Standorte ab und wird vom Klima, von der Kontinentalität und der nördlichen Breite des Gebietes beeinflusst. Die Diversität der Vegetation wird durch den Shannon-Index, den Simpson-Index und die Q-Statistik für die „Arten/Präsenz-Verteilung" geschätzt. Die Werte sind in den untersuchten Berggebieten sehr ähnlich.

Fritz Hans Schwarzenbach, Kistlerweg 9, CH-3006 Bern

1. Introduction

1.1 An old problem has found a new interest

The altitude zonation of the vegetation in alpine regions is a well-known phenomenon and has been studied in the Alps since the early times of scientific botany. However, in spite of the rich literature there are still many questions to be answered.

The world-wide discussion of the so-called "greenhouse-effect" with its assumed changes of the climate has inspired new projects. Many scientists are studying the possible influence of rising temperatures, changing winds and precipitation on the future development of the vegetation at high altitudes (e.g. Graham & Grimm 1990, Holt 1990, Huntley 1991, Körner 1994, Grabherr et al. 1994, 1995, Moore 1990, Theurillat 1995). It is assumed that the alpine vegetation moves to higher altitudes with increasing temperatures (Gottfried et al. 1994).

1.2 Analysis of the problem

The present altitude distribution of vascular plants in the mountains of Greenland is the result of a development over time of the present ecosystems, influenced by interactions of biological and environmental factors. This evolutionary process is characterised by multiple changes in environmental conditions influencing the immigration, the persistence and the selection of species now found in Greenland.

The results of recent studies show that Greenland has been influenced by strong changes of climate and glaciation during the Pleistocene (e.g. Funder 1989 a, 1989 b, Funder & Abrahamsen 1988, Funder & Hjort 1973). As a consequence, the altitude distribution of plants is the result of a development of the ecosystems since then. The present pattern of the vertical distribution can only be understood if the history of the vegetation is considered, and if the dynamics of the evolutionary processes of arctic ecosystems are analysed under many different aspects. The analysis of the long term development of ecosystems beyond the treeline is a difficult task. Thus, when dealing with the complex problem of the altitude distribution it is necessary to find an appropriate approach and to develop and test some new methods. Many ecological problems look very simple but are most difficult to solve. As a rule, the analysis of such problems requires a multidisciplinary approach.

Human activities are believed not to have influenced directly on the vegetation in the mountains of East and Northeast Greenland, and, therefore, these areas provide an opportunity to study a flora undisturbed by human interference.

1.3 History of the project

The project started in the summer of 1948, when the author assisted the geologist Dr H. Bütler during the Danish East Greenland Expedition lead by Dr Lauge Koch. It was a six-weeks trek around and through the Giesecke Bjerge (Gauss Halvø) using Icelandic ponies to carry equipment and food. The following expeditions visited North Andrée Land (1949), South Andrée Land (1950), Stauning Alper, Nathorst Land (1951, 1954), Ole Rømer Land, and Kronprins Christian Land (1952). In those days geologists of the Danish East Greenland Expeditions explored geologically unknown regions in teams of 2-3 men. They reached their base camps by motorboat or seaplane (Fig. 1). They studied the geology during long excursions starting either from base-camp at seashore or using transit camps at higher levels which they established by carrying all equipment and food on their backs. By 1952, access and working methods improved using seaplanes for geological reconnaissance flights and aerial photography as well as for transportation to open lakes in the interior. So, in the summer of 1952, two teams were transported by a Norseman seaplane on a reconnaissance trip to Vibeke Sø (Ole Rømer Land) and later by a Catalina seaplane to Centrum Sø in Kronprins Christian Land. The first heli-

Fig. 1. Ella Ø (1956).

copter was chartered in 1956, when the team of Dr J. Haller had the chance to study the geology of the nunataks north of 74°N.

The botanical fieldwork was carried out during excursions planned by the geologists. This arrangement had the advantage that plants were observed and collected at inland sites and at different altitudes. On the other hand, the scientific equipment had to be reduced to a minimum and the botanical work had to be based on simple and time saving methods.

The fieldwork in mountains of East and North Greenland has continued in later years in southern Werner Bjerge, (71°N, 1991), in Gåseland (70°N, 1994), in North Peary Land (83°N, 1995), and in Zackenberg (74°N, 1998). These areas have been reached by Twin Otter aircrafts. The fieldwork was done in the traditional way by botanical excursions starting from base-camp at sea level or from a transit camp at higher altitude.

2. Selection and description of the 5 study areas (called WA1-WA5)

2.1 Criteria for selection

The following criteria have been used to select the study areas:

- Geographically well defined area.
- Areas with northern latitudes between 71° and 81° representing coastal and inland areas (with nunataks).
- Similar climatic conditions within each individual area.
- Sufficient numbers of observations and herbarium specimens for data analysis.
- The vegetation line could be reached by excursions.

Only five of the investigated areas fulfil these criteria (WA1-WA5) (Fig. 2).

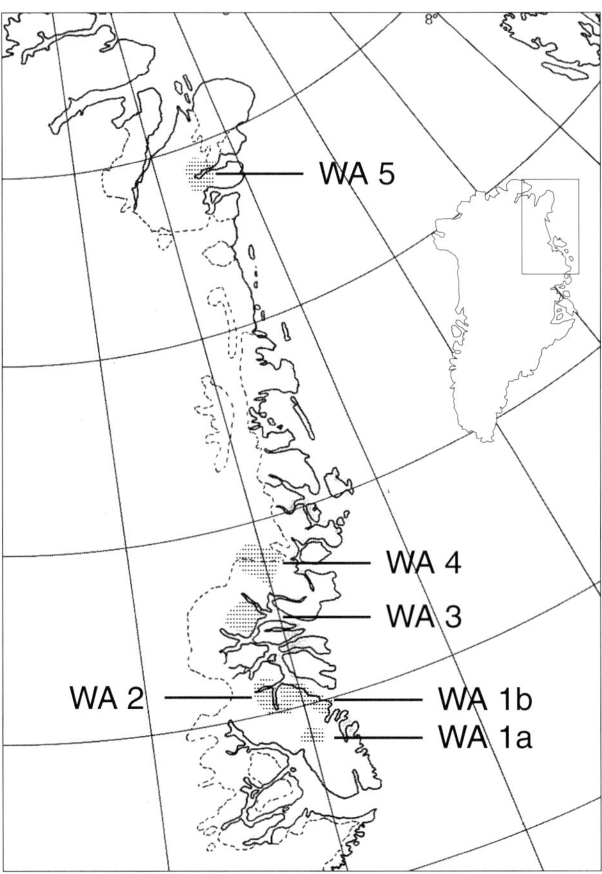

Fig. 2. Map of Greenland showing the study areas WA1a, WA1 b, WA2-WA5.

2.2 Study area: Southern Werner Bjerge and eastern Stauning Alper WA1 (Fig. 3)

2.2.1 Southern Werner Bjerge (sub-area) (Fig. 4)

71°46′-71°51′N, 24°02′-24°19′W, altitude range: 80 – 1680 m
Schuchert Dal, Pingo Pass, Pingo Dal, Biskop Alf Dal, mountains south of Randspids.
20.07.91 – 17.08.1991, E. Schwarzenbach, F. H. Schwarzenbach

Landscape

"Pingo Pass" c. 500 m a.s.l. (Fig. 5), leads from Pingo Dal in the East to the Schuchert Dal in the West separating the alkaline massif of the Werner Bjerge in the north from the sedimentary part of northern Jameson Land (Henriksen et al.1980).

Geology

The geology of Jameson Land and of the Werner Bjerge is well known (Stauber 1940, Bierther 1941, Witzig 1954, Bearth 1959, Kempter 1961, Defretin-Lefranc et al. 1969, Callomon et al. 1972, Perch-Nielsen et. al. 1975). A geological map 1:100000 is found in Henriksen et al. (1980).

The Werner Bjerge are a Tertiary intrusive complex (Bearth 1959, Henriksen et al. 1980). The mountains just north of „Pingo Pass" (rising to c. 1700 m) are built up from palaeozoic and lower Triassic sediments. The same kind of sediments are found to the south where the chain of the Gurreholm Bjerge flanks

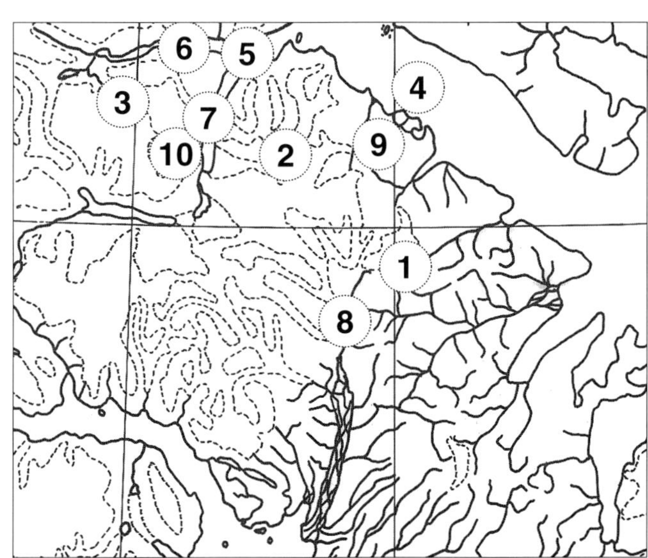

Fig. 3. Map of study areas WA1 a, WA1 b, WA2: (1) Southern Werner Bjerge, (2) Stauning Alper, (3) Nathorst Land, (4) Kong Oscar Fjord, (5) Segelsällskapet Fjord, (6) Forsblad Fjord, (7) Alpefjord, (8) Schuchert Dal, (9) Skeldal, (10) Schaffhauserdal.

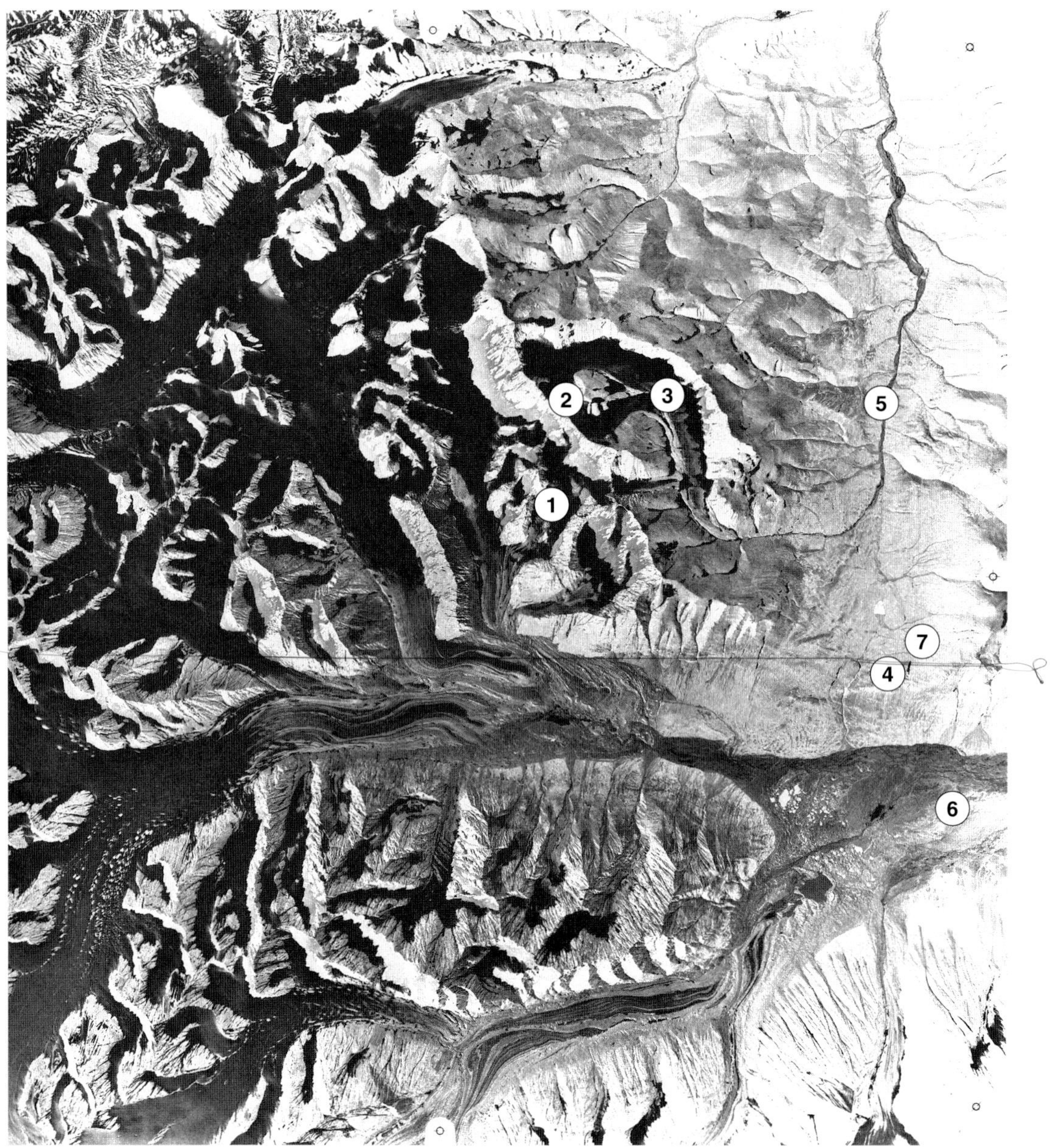

Fig. 4. Aerial photo of WA1 a, Southern Werner Bjerge: (1) Southern Werner Bjerge, (2) Randspids, (3) Biskop Alf Gletscher, (4) Pingo Pass, (5) Pingo Dal, (6) Schuchert Dal, (7) Lomsø. [AF 87.716 MM, 888-M-239, reproduced with permission of Kort- og Matrikelstyrelsen, København, Danmark].

the eastern side of Schuchert Dal. A few outcrops of diarists/grano-diorites have been mapped at the western edge of the plateau and further on to the slope. Some basic dikes and sills have been observed (Henriksen et al. 1980).

The Quaternary geology of East Greenland between latitudes 70°N and 75°N has been studied further by Funder & Hjort (1973), Funder & Abra-

Fig. 5. Base camp near the airstrip near Pingo Pass. View towards the southern Stauning Alper west of upper Schuchert Dal (WA1 a, 1991).

hamsen (1988), Funder (1989 a, b). There are many local advances and retreats of glaciers in the Holocene Period. Glacial and alluvial deposits cover large parts of the study area. Pingo Pass is bordered by a series of moraines left by the ice which spilt over from the Schuchert Dal to flow down Pingo Dal and which has resulted in the formation of the small lake Lomsø just east of the pass. Many big erratic boulders are found on the plateau. A conspicuous old moraine of a former advance of the Biskop Alf Gletscher marks the eastern end of the Pingo Pass plateau (Fig. 4).

Climate

The climate of the area is comparable with that of Jameson Land, which is fairly continental. The nearest station is Station 133, 71°10′ N, 23°37′ W, 261 a.s.l. (Anonymous 1989). Station Mesters Vig (72°15′ N, 23°54′ W) has a cooler, more coastal type climate (Tab. 1).

The temperatures of the Pingo Pass area may be approximately 1°C lower than at Station 133 due to the higher altitude. Data for the monthly precipitation are published only for Station Mesters Vig, 1961/85 (Ohmura and Reeh, 1991). They show that the total annual precipitation in the area ranges between 300 mm and 400 mm, falling mainly as snow.

Table 1. Mean monthly temperatures and precipitations (Station 133, Mesters Vig)

	May	Jun	Jul	Aug	Sep
Mean temp. (°C)					
Station 133 (1982-1993)	–1.5	6.2	9.3	7.6	0.3
Mesters Vig (1961-1985)	–4.8	1.6	5.2	5.3	–0.6
Precipitation (mm)					
Mesters Vig	13.8	18.3	28.5	22.7	29.3

Vegetation

The number of species in the study area is not as high as in the adjacent parts of Jameson Land which is surprisingly rich and includes taxa otherwise rare in East Grenland (Bay & Holt 1986, Fredskild et al. 1982, 1986). The smaller number of species in WA1, Southern Werner Bjerge, might be explained in part by the higher altitude (most sites are above 400 m) and in part by the low habitat diversity.

The uppermost Schuchert Dal is exposed to dry winds blowing from the glacial area of the southern Stauning Alper. Along the west-facing slope of the valley (up to an altitude of about 400 m) there are snowbeds with *Betula nana*, *Vaccinium uliginosum* ssp. *microphyllum* or *Dryas octopetala s.l.* (see chapter 3 , Taxonomic considerations). On dry and wind-swept terraces along the river a fragmentary "föhn-steppe" has been observed with *Calamagrostis purpurascens, Carex nardina, C. rupestris, Kobresia myosuroides, Braya purpurascens, B. humilis,* and *Lesquerella arctica*. Mires are restricted to banks and the deltas of the side rivers.

The rather flat plateau of Pingo Pass between 400-700 m a.s.l. (Fig. 5) is covered mainly by a "hummocky heath". *Cassiope tetragona* prefers moderately moist grooves between the hummocks. The *Dryas octopetala s.l. – Carex nardina*-heath grows on dry terraces. Wet heaths with *Eriophorum triste, E. scheuchzeri,* and *Carex bigelowii s.l.* form the transition to mires along the rivers and in shallow depressions. An unusual type of moist heath dominated by *Salix arctica* and *Polygonum viviparum* has been found along the rivers at the foot of the mountains to the south. Different types of snow-bed vegetation are well developed below persisting snow drifts.

The north-facing slopes of the mountains south of Pingo Pass have a poor fell-field vegetation on large screes with blocks and shales.

Herb-slopes are rare in this area, they are restricted to a few sunny places near the western side moraine of the Biskop Alf Gletscher at an altitude of 700-770 m. The commonest species are *Arnica angustifolia, Antennaria canescens, Taraxacum brachyceras, Sibbaldia procumbens, Thalictrum alpinum, Botrychium lunaria, Festuca rubra s.l., Potentilla crantzii, Luzula spicata,* and *Gentiana tenella.*

The steep slopes of the mountains north of Pingo Pass above 800 m are characterised by fell-field vegetation and in wet parts by snow-patch vegetation. A remarkable solifluction field (1000 m) has a high number of species among them *Eriophorum triste, Colpodium vahlianum,* and *Cochlearia groenlandica.* A well developed snow-patch facing to East was found at 1250 m with *Colpodium vahlianum, Phippsia algida s.l.,* and *Carex maritima.*

A rather uniform heath dominated mostly by *Cassiope tetragona* was found on the terraces at the southern side of Pingo Dal. The altitude limit of *Betula nana* lies at only 350 m due to the northern exposure. On the western shoulder of the southernmost side valley of Pingo Dal at altitudes between 660 m and 840 m grew a more or less closed cover of *Salix arctica* with *Polygonum viviparum* as a codominant species. Below the fronts of solifluction streams (650 m) two herb-slopes with *Poa alpina s.l., Taraxacum brachyceras, Potentilla crantzii, Thalictrum alpinum,* and *Gentiana tenella* have been found.

The airstrip and associated dirt roads (constructed 1957) have been recolonised by about 2/3 of all species of the Pingo Pass area as observed in 1991 (Schwarzenbach 1996a).

2.2.2 Eastern Stauning Alper (sub-area)

72°12'-72°25'N, 24°14'-24°37'W, altitude range: 0-730 m (Fig. 6)
Skeldal, Bersærkerbræ, Flødegletscher, Syltoppene, Kap Peterséns, lowermost Skjoldungebræ.
21.07./26.07.1951 and 16.08./01.09.1951, E. Fränkl, P. Braun, F. H. Schwarzenbach

Landscape and geology

The Syltoppene (Figs. 7, 8) is a chain of steep mountains with altitudes up to nearly 1500 m with limestone, dolomites, and tillites of the Precambrian Eleonore Bay Formation (Fränkl 1953 a, Haller 1958, 1970). The summits are intersected by rather steep valleys with small glaciers, moraines, and screes. The coastal belt is rather narrow. The stony and often dry terraces create only poor conditions for the vegetation.

Skeldal is a broad and flat valley running from SSW to NNE. The big river is flanked by conspicuous terraces and merges with a big delta into Kong Oscar Fjord. The middle part of the valley (below the end-moraine of Bersærkerbræ) is sheltered by mountains and seems to be rather dry. There are enormous accumulations of quaternary deposits in the main valley (Fig. 9).

The Flødegletscher and the northern tributary glacier of the Bersærkerbræ originate in cirques of an outer chain of the Stauning Alper. Both glaciers represent the alpine type of glaciers to be found throughout the massif of the Stauning Alper. The rocks on both sides of the two valleys form steep walls.

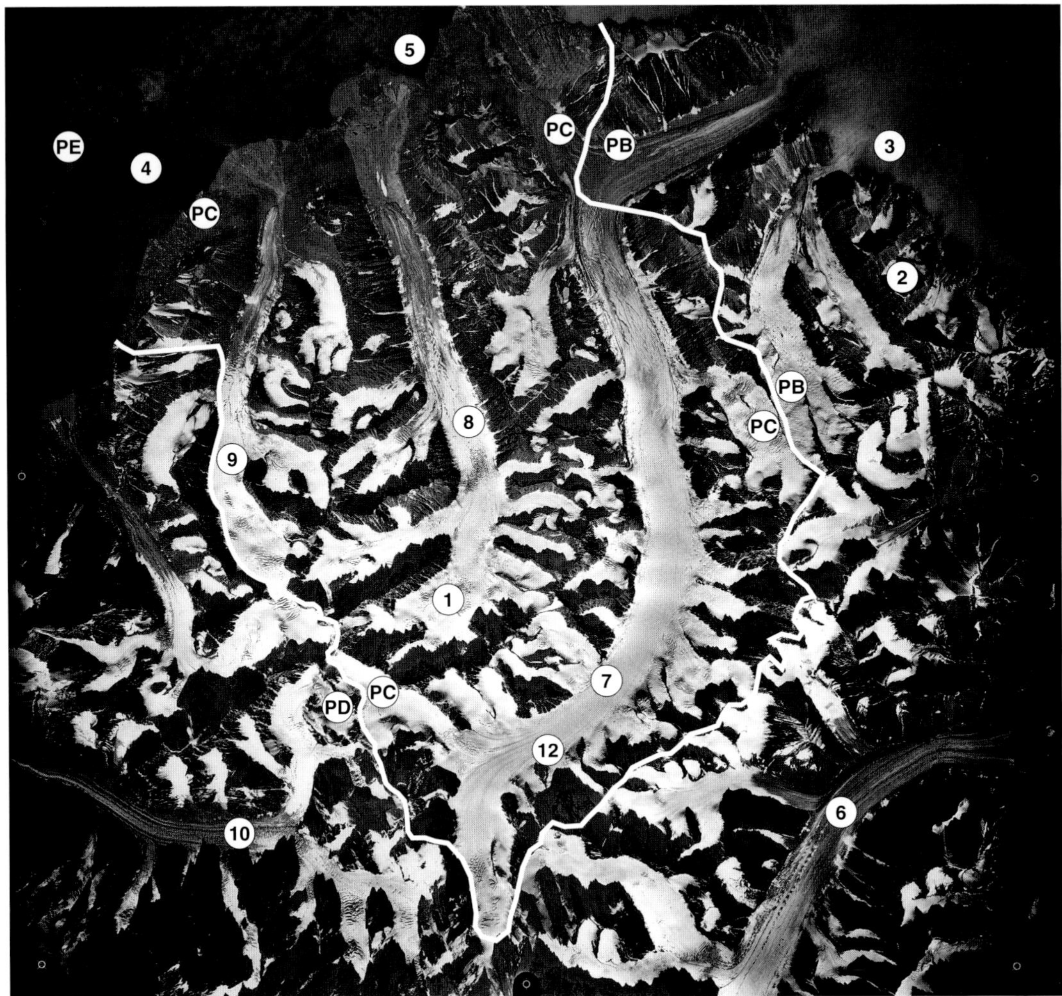

Fig. 6. Aerial photo of WA1 b and part of WA2, Stauning Alper: (1) Stauning Alper, (2) Syltoppene, (3) Kong Oscar Fjord, (4) Alpefjord, (5) Segelsällskapet Fjord, (6) Bersærkerbræ, (7) Skjoldungebræ, (8) Linné Gletscher, (9) Sedgwick Gletscher, (10) Vikingebræ, (11) Skeldal, (12) Elisabeth Bjerg, (PB, PC, PD, PE) sub-areas called ‚provinces'. [AF 87.72 MM, 888-L-3866, reproduced with permission of Kort- og Matrikelstyrelsen, København, Danmark].

Climate

The climate of the western coast of Kong Oscar Fjord is quite similar to the Mesters Vig area (Tab.1). The average monthly temperatures at Mesters Vig have positive values only from June to August (+1.6°C, + 5.2°C, +5.3°C). The summer temperature is rather low due to the influence of the fjord creating fog. Inversions of temperature have been observed in the Syltoppene at levels of several hundred meters above sea. The temperature in the middle part of Skeldal is higher.

The amount of precipitation for the five months May to September based on the monthly average for the period 1961/85 in Mesters Vig reaches 112.6

Fig. 7. Syltoppene, glacier flowing eastwards to Kong Oscar Fjord. The mountains of Traill Ø in the background. (WA1 b, 1951).

Fig. 8. The uppermost part of Skjoldungebræ with the western side of Syltoppene seen from top of Elisabeth Bjerg (WA1 b, WA2, 1951).

Fig. 9. Camp near Bersærkerbræ (WA1 b, 1951).

mm with the highest values in July and September (Tab. 1). Sea fog entering Kong Oscar Fjord is quite common during the summer.

Vegetation

The vegetation is rather poor along the slope between Syltoppene and Kong Oscar Fjord due to the unfavourable conditions of soils and climate. The middle and the lower part of the Skeldal offer a broad variety of habitats including some rare species such as *Lycopodium annotinum, Thalictrum alpinum, Harrimanella hypnoides* and *Carex bicolor*. The dry heath with *Betula nana, Arctostaphylos alpina, Empetrum nigrum* var. *hermaphroditum* and *Rhododendron lapponicum* is well represented on terraces.

The altitude limit of vascular plants is reached at 1450 m in the Syltoppene. This altitude is similar to that in the mountains north of the Pingo Pass (1460 m). This rather low altitude limit on Syltoppene is explained by the maritime influence of the Kong Oscar Fjord. In the Pingo Pass area the altitude level is influenced by the cold and rather wet winds blowing from the East, which seem to lower the summer temperature considerably. This similar level of altitude limit of vascular plants is one of the reasons for combining the two geographically separated sub-areas in WA1.

2.3 Study area WA2: Northern and western Stauning Alper, Nathorst Land (Fig. 6)

72°01′-72°28′N; 24°38′-26°48′W; altitude range: 0-2930 m

The area WA2 includes the northern and western Stauning Alper together with Nathorst Land. There are several regions which are listed below:

- Vikingebræ, Alpefjord: 26.08./31.08.1950, E. Fränkl, P. Braun, F. H. Schwarzenbach
- Northern Skjoldungebræ: 27.07./15.08.1951, E. Fränkl, P. Braun, F. H. Schwarzenbach
- Linné Gletscher, Sedgwick Gletscher: 27.07./15.08.1951, E. Fränkl, P. Braun, F. H. Schwarzenbach
- Linné Gletscher: 26.08./01.09.1954, J. Haller, W. Diehl, F. H. Schwarzenbach
- Vikingebræ: 26.08./30.08.1950, E. Fränkl, P. Braun, F. H. Schwarzenbach
- Vikingebræ, Gullygletscher, Sefstrøm Gletscher: 30.07.-25.08.1954, J. Haller, W. Diehl, F. H. Schwarzenbach
- Schaffhauserdalen: 24.08.-30.08.1952, E. Wenk, H. Buess, O. Wackernagel, K. Stucky, F. H. Schwarzenbach
- Forsblad Fjord, Højedal, Tærskeldal, Ismarken: 24.07.-29.07.1954, J. Haller, W. Diehl, F. H. Schwarzenbach

2.3.1 Landscape, geology, climate

Landscape

The Stauning Alper separate the Palaeozoic and Mesozoic sediments of Jameson Land in the south and the Precambrian series in the north. They form an important climatic barrier between the maritime conditions of Kong Oscar Fjord which are similar to the outer coast, and the continental climate in western Nathorst Land and Lyell Land. It is reasonable to consider the Stauning Alper also as a phytogeographical barrier between the Scoresby Sund area in the south and the fjord system of the Kejser Franz Joseph Fjord in the north. The area of the northern and western Stauning Alper is shown on an aerial photo, the northern part marked by "C", the western part by "D" (Fig. 6).

The Stauning Alper are a mountainous area of about 2000 km^2 (Fig. 10). The mountains are very steep and have been visited frequently by mountaineering expeditions (Fantin 1969). The highest mountain, Dansketinden, reaches an altitude of more than 2900 m. The valleys have glaciers. The cirques of the glaciers lie between 1200 m and 1500 m.

Skjoldungebræ (Figs. 8, 10, 11), Linné Gletscher and Sedgwick Gletscher

Fig. 10. View from Silberhorn towards the central part of Stauning Alper with Frihedstinde (highest summit to the right) and the innermost side valleys of Linné Gletscher (WA2, 1951).

Fig. 11. Elisabeth Bjerg from the firn basin of the uppermost Skjoldungebræ (WA2, 1951).

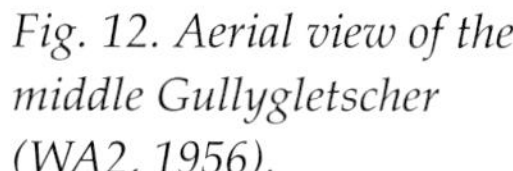
Fig. 12. Aerial view of the middle Gullygletscher (WA2, 1956).

Fig. 13. Uppermost camp at the northern side of Gullygletscher (WA2, 1954).

Fig. 14. Aerial view of Sefstrøm Gletscher from north (WA2, 1956).

flow northwards to Segelsällskapet Fjord. Vikingebræ, Gullygletscher and Sefstrøm Gletscher (Figs. 12, 13, 14) flow to Alpefjord in the west. The double glacier "Gullygletscher/Sefstrøm Gletscher" reaches the shore of the fjord.

Schaffhauserdalen on the western side of Alpefjord leads far into Nathorst Land. The mountains on both sides of the valley are steep. The slopes are covered by loose scree of highly metamorphic rocks.

A steep slope leads from the end of Forsblad Fjord to the Tærskeldal (about 700 m) with a few small lakes surrounded by ice-polished rocks and active polygon-fields. Højedal (Fig. 15) in the north provides a possible route to the edge of Ismarken, a rather big icecap (highest point 2010 m).

Geology

E. Fränkl was the first geologist to study the Precambrian sediments in the northern and eastern parts of the Stauning Alper (Fränkl 1951, 1953 a). He was followed by J. Haller who studied the so-called "Der Zentrale Metamorphe Komplex" in the Stauning Alper and around Forsblad Fjord (Haller 1958, 1970).

Climate

The nearest weather station is Mesters Vig (Tab. 1). However, the data show

Fig. 15. Nathorst Land, Højedal (WA2, 1954).

Fig. 16. Nathorst Land, Spærregletscher. Cloud of drifting sand (WA2, 01.08.1956).

the conditions at the coast of Kong Oscar Fjord and are not representative for the central and western part of the Stauning Alper. Measurements in the field have shown that summer temperatures in the interior of the valleys are considerably higher than that at MestersVig. Based on the altitude limits of the commonest species of vascular plants there could be a difference of the mean summer temperature of about 1.5°C. This is because the valleys are fairly sheltered from winds blowing sometimes strongly on the summits and ridges. Fog and temperature inversions are rare. Compared with the Syltoppene precipitations seem to be much lower.

The Alpefjord as well as the Forsblad Fjord and the Segelsällskapet Fjord are exposed to violent föhn gales blowing from the Inland Ice (Fig. 16). The föhn wind may raise the temperature sometimes considerably. This was observed on the 28/08/1954, when 23°C was measured in the base camp at the head of Forsblad Fjord.

The continental climate of these inner fjords is also reflected by the well developed sites of the "föhn-steppe" at high levels in Højedal near Ismarken.

2.3.2 Botanical exploration and vegetation

Botanical exploration

The study area WA2 has been visited by botanists twice before 1930. Sørensen (1933) refers to two localities only: Polhelm Dal, Lyell Land, about 72°27'N, 26°10'W (Hartz & Kruuse 1911) and Head of Forsblad Fjord, about 72°27'N, 26°20'W (Dusén 1901, Hartz & Kruuse 1911). Schwarzenbach studied the mountain vegetation during the Danish East Greenland Expeditions 1950, 1951, 1952 and 1954. Since then, botanists and mountaineers have added to collections and these have been studied by Dr G. Halliday.

Vegetation

The gently rising slopes at the southern coast of Segelsällskapet Fjord with numerous terraces are covered by a dry to wet heath mostly dominated by *Cassiope tetragona*. *Betula nana* is rather rare. *Eriophorum* spp. and *Carex* spp. are common in the mires at low level.

Typical fell-field vegetation is found in the interior parts of Skjoldungebræ, Linné Gletscher and Sedgwick Gletscher and in the adjacent side valleys up to levels of about 1800 m. Remarkable species are *Carex norvegica* (920 m, 1030 m, 1120 m), *Carex glacialis* (1030 m, 1300 m), and *Erigeron compositus* (1150 m) found along Linné Gletscher. The altitude limit of vascular plants reaches its highest level at 2200 m: A flowering plant of *Saxifraga cernua* was found near the summit of Elisabeth Bjerg (Figs. 8, 11).

The fell-field is also well developed along Vikingebræ, Gullygletscher and Sefstrøm Gletscher. Surprisingly rich collections have been made at the foot of south- or west-facing rocks behind the lateral moraines at levels up to 1600 m along Sefstrøm Gletscher. Fragments of southern plant communities have been found in sheltered niches (Fig. 42). At an altitude of 1230 m *Arnica angustifolia, Calamagrostis purpurascens, Euphrasia frigida, Hierochloë alpina, Pedicularis flammea,* and *Taraxacum arcticum* were growing together. At a level of 1030 m *Erigeron humilis, Harrimanella hypnoides, Huperzia selago, Minuartia biflora, Poa arctica, Salix herbacea, Sibbaldia procumbens, Taraxacum brachyceras* and *Tofieldia pusilla* have been found in a snow-patch. This group of species is similar to a type of snowbed vegetation found in Hjørnedal, Gåseland (70°N) at about 700 m (Schwarzenbach, unpublished).

On the south-facing slope north of Vikingebræ a fragment of a southern herb-slope was observed. Recorded are *Antennaria canescens, Arnica angustifolia, Campanula gieseckiana s.l., Euphrasia frigida, Festuca rubra s.l., Luzula spicata, Poa alpina s.l., Potentilla crantzii, P. hyparctica, Sibbaldia procumbens, Taraxacum brachyceras,* and *Viscaria alpina* among other species. *Arabis holboellii* and *Draba aurea* have been collected at 360 m.

It was possible to study the vertical distribution of vascular plants on a south-facing and on a north-facing profile of Schaffhauserdalen from the bottom of the valley to the altitude limit of vascular plants with the highest observation at 2050 m.

The rather flat coastal area at the end of Forsblad Fjord has quite a rich vegetation in contrast to Tærskeldal (about 700 m). The bottom of this valley is stony and offers poor conditions. In the valley Højedal in a WSW direction plants are growing right up to the edge of the flat ice-shield of Ismarken (1600 m). Species of dry sunny slopes and of the continental "föhn-steppe" were present at several places. *Carex supina* ssp. *spaniocarpa* was collected at a level of 1100 m. *Hierochloë alpina, Melandrium triflorum* and *Potentilla hookeriana s.l.* reached the altitude of 945 m. *Calamagrostis purpurascens, Carex nardina, C. rupestris, Draba arctica,* and *Dryas octopetala s.l.* were found at 860 m.

2.4 Study area WA3: East Andrée Land (Fig. 17)

73°15′-73°46′N; 25°02′-26°31′W; altitude range: 0-2150 m

- Junctiondal, Månesletten, Gauligletscher, Benjamin Dal, Luciadal, Blåbærdal. 08.07.-07.08.1950, 15.08.-24.08.1950, E. Fränkl, F. H. Schwarzenbach
- Morænedal, Endeløs Gletscher, Nøkkefossen, Agardh Bjerg. 28.07.-25.08. 1949, E. Fränkl, F. H. Schwarzenbach
- Grejsdal, Kap Weber. 09.08.-13.08.1950, E. Fränkl, P. Adams, F. H. Schwarzenbach

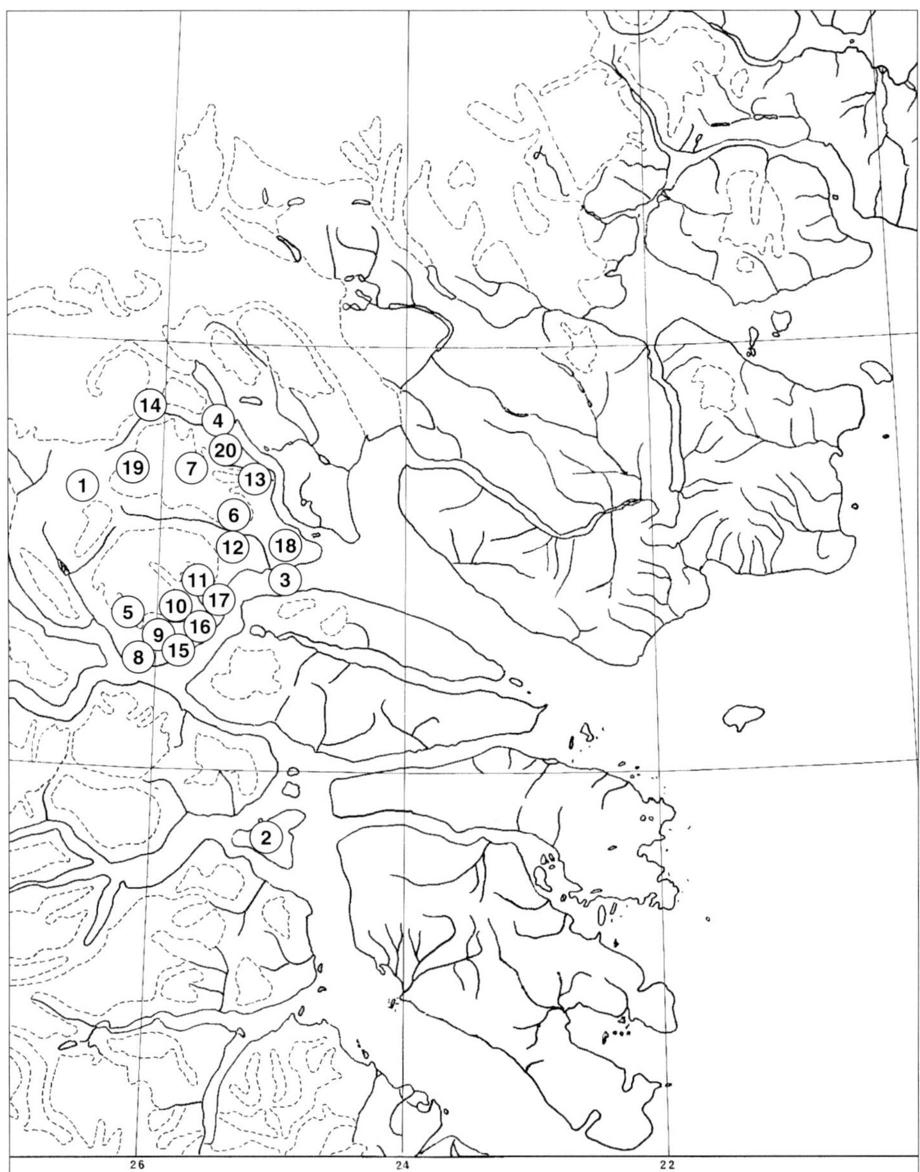

Fig. 17. Map of WA3, Andrée Land: (1) Andrée Land, (2) Ella Ø, (3) Kejser Franz Joseph Fjord, (4) Geologfjord, (5) South Andrée Land, (6) North East Andrée Land, (7) North Andrée Land, (8) Junctiondal, (9) Benjamin Dal, (10) Lucia Dal, (11) Blåbærdal, (12) Grejsdal, (13) Morænedal, (14) Eremitdal, (15) Plateau east of Junctiondal, (16) Hvidevæggen, (17) Teufelsschloss, (18) Kap Weber, (19) Længselbjerg, (20) Agardh Bjerg.

2.4.1 Landscape, geology, climate

Landscape

The area is situated between Kejser Franz Joseph Fjord in the east and the two valleys Grejsdal and Eremitdal in the west. In the north the Geologfjord separates Andrée Land from Strindberg Land. East Andrée Land might be described as a huge plateau rising from about 1300 m in the east to more than 2100 m in the west (Fig. 18). The western part is covered partially by ice-caps. The valley Grejsdal separates the southern and the northern half of East Andrée Land.

The southern plateau (Månesletten) drops with steep slopes towards the fjords (Fig. 19). The plateau is best reached by ascending through the narrow Junctiondal. In the east the broad and rather flat Benjamin Dal (Fig. 20) behind the impressive Teufelsschloss (1410 m) leads to the interior valleys of Blåbærdal and Luciadal.

The northern part of the study area includes Kap Weber, north of Grejsdal (marked 1230 m) and the Morænedal with its side-valleys (Fig. 21). The highest mountain is Længselbjerg (about 2030 m). The two glaciers Endeløs Gletscher and Døde Slette have their end-moraines in the lower Moræne-dal.

Fig. 18. Aerial view of Antarctic Sund and Hvidevæggen (WA3, South Andrée Land, 1956).

Fig. 19. Aerial view of Gauli Gletscher, Månesletten, and plateau east of Junctiondal (WA3, South Andrée Land, 1956).

Geology

The geology of the area has been studied by Fränkl (1953 b) and (in parts) by Haller (1953). Koch and Haller (1971) have published a geological map 1:250000. The major part of the area is dominated by the Precambrian sediments of the Eleonore Bay formation (Fig. 21). An interesting geological feature is the Grejsdalen Granite Batholith. A beautiful example of a double fold from the late Caledonian can be seen in the Blåbærdal (Fig. 22). The so-called "Kalk/Dolomit-Series" is well exposed on Månesletten in the southern part of the area and offers the opportunity to study the vegetation on calcareous and dolomitic soils near the altitude limit of vascular plants.

Climate

The nearest weather station is Ella Ø (72°50'N, Fig. 1). According to Gartmann (1993) the mean temperature for July at Ella Ø is +9.4° C. This level of summer temperature seems to be similar to that in East Andrée Land based on recordings in the field (Schwarzenbach, unpubl.). The annual precipitation is between 200 and 300 mm (Ohmura and Reeh 1991).

Fig. 20. Aerial view of Teufelsschloss from west (WA3, South Andrée Land, 1956).

Fig. 21. North Andrée Land: Moræenedal (WA3, 1949).

Fig. 22. Aerial view of the folds in Blåbærdal (WA3, South Andree Land, 1956).

Benjamin Dal, Luciadal and Blåbærdal are sheltered and rather calm valleys, while the Moraenedal and the Geologfjord are extremely exposed to föhn storms blowing from the west.

2.4.2 Botanical exploration and vegetation

Botanical exploration

The first important paper was published by Vaage (1932). One year later Th. Sørensen, (wintering 1931/32 at Ella Ø) published his comprehensive paper on the vascular plants of East Greenland between 71°00'N- 73°30'N (Sørensen 1933). He studied four areas (Traill Ø, Ymer Ø, Ella Ø and Kap Hedlund in Lyell Land) and included a few scattered places at Kempe Fjord and Rhoess Fjord. Summarizing the botanical investigations from earlier expeditions he has listed 43 localities, which omits the collections from Copeland & Pansch of the "Die Zweite Deutsche Polar-Expedition 1869-1870" in Buchenau & Focke [1874] on account of the lack of precise topographic positioning of the localities. He mentions two places in the study area:

- WA3: Junctiondal (73°14′N, 25°52′W) visited by G. Seidenfaden
- WA3: Renbugt (ca. 73°22′N, 26°38′W) visited by J. Vaage (1932).

Gelting (1934) adds three more places form the innermost Geologfjord:

- WA3: Primulabugt (73°49′N, 25°20′W) visited by G. Seidenfaden (18.08.1929)
- WA3: Nunatak Gletscher (73°55′N, 25°48′W) visited by P. Gelting (09.11.1931)
- WA3: Eremitdal (73°49′N, 25°33′W) visited by P. Gelting (10/11/1931).

Dr G. Halliday is preparing distribution maps of all vascular plants including the Kejser Franz Joseph Fjord area. These maps are based on the specimens in the Greenland Herbarium København, which include the author's collections as well as the plants found by other expeditions in recent years.

Vegetation

An interesting type of "föhn-steppe" is represented at high levels up to 1285 m on calcareous and dolomitic soils of the plateau "Månesletten". Typical species are *Calamagrostis purpurascens, Braya purpurascens, B. linearis, Draba bellii, Melandrium affine, Minuartia rubella, Potentilla nivea. Potentilla hookeriana s.l., Erigeron eriocephalus, Taraxacum phymatocarpum, Festuca baffinensis, Carex nardina, Poa abbreviata, Poa glauca,* and – strange enough – *Poa pratensis s.l.*

In snow-patches on the high-plateau of Månesletten (around 1200 m) *Colpodium vahlianum, Draba alpina s.l., D. adamsii, D. nivalis, Juncus biglumis, Phippsia algida s.l., Ranunculus affinis, Saxifraga rivularis s.l.,* and *Taraxacum arcticum* have been found.

A conspicuous geomorphological feature are the terraces flanking the river in the lowest parts of Benjamin Dal covered by a dry and sometimes rather open heath (Schwarzenbach 1951, 1960). Common species are *Betula nana, Cassiope tetragona, Dryas octopetala s.l., Rhododendron lapponicum,* and *Vaccinium uliginosum* ssp. *microphyllum* besides *Carex scirpoidea, Pedicularis flammea,* and *Pyrola grandiflora.*

In the lower part of Moraenedal a similar type of a dry heath with *Dryas octopetala s.l., D. integrifolia s.str., Cassiope tetragona, Arctostaphylos alpina* (380 m), *Empetrum nigrum, Rhododendron lapponicum* and *Vaccinium uliginosum* ssp. *microphyllum* is common. *Hippuris vulgaris* has been found in a pond at 240 m. There is a well developed fell-field vegetation at the south-facing slope of Agardh Bjerg (upper Moraenedal) between 1250 m and 1500 m with *Arenaria pseudofrigida, Braya purpurascens, Campanula uniflora, Cardamine bellidifolia, Carex nardina, Cerastium arcticum s.l., Draba bellii, Hierochloë alpina, Luzula confusa, Papaver radicatum s.l., Phippsia algida s.l., Potentilla hyparctica, Saxifraga caespitosa, S. cernua,* and *S. nivalis.* Unusual findings at this altitude are *Carex*

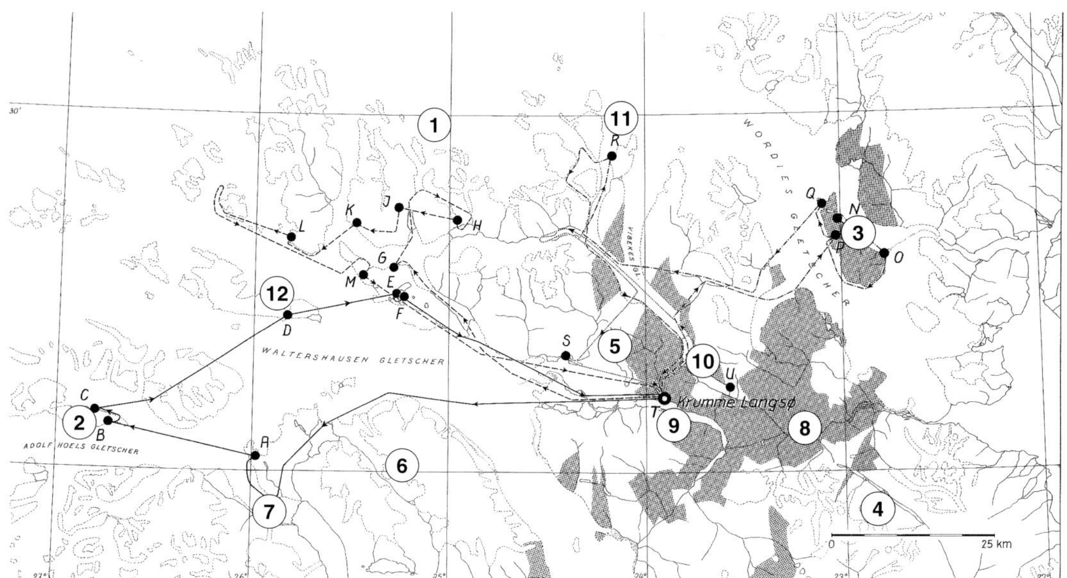

Fig. 23. Map of WA4: (1) Nunataks north of 74°N, (2) Bernhard Studer Land, (3) C.H. Ostenfeld Nunatak, (4) Hudson Land, (5) Ole Rømer Land, (6) Strindberg Land, (7) North Andrée Land, (8) Promenadedal, (9) Krumme Langsø, (10) Vibeke Sø, (11) Vibeke Nunatak, (12) Waltershausen Nunatak, (A-U) Landing points during the helicopter flight of 1956 (see Schwarzenbach 1961).

supina ssp. *spaniocarpa* (1140 m), *Arnica angustifolia* (1050 m), *Cassiope tetragona* (1220 m), and *Ranunculus glacialis* (1100 m).

2.5 Study area WA4: Nunataks north of 74° N (Fig. 23)

74°02'-74°27'N; 22°02'-26°50'W; altitude range: 200-1470 m

- Vibeke Sø (reached by sea-plane), Promenadedal. 6.07.-29.07.1952, E. Fränkl, F. Müller, F. H. Schwarzenbach
- Krumme Langsø (reached by sea-plane), nunataks between Adolf Hoel Gletscher and C. H. Ostenfeld Nunatak (landings by helicopter). 14.08./19.08.1956, J. Haller, F. H. Schwarzenbach (Fig. 24)

2.5.1 Landscape, geology, climate

Landscape

The area WA4 covers the area of mountains and nunataks (Fig. 25) between Hudson Land in the east and the Inland Ice in the west. The southern border of the area runs from Vibeke Sø and Promenadedal (Hudson Land, Fig. 26) in

Fig. 24: Helicopter at the landing place B in Bernhard Studer Land (WA4, 1956).

Fig. 25. Aerial view of the small lakes on the C.H. Ostenfeld Nunatak (WA4, 1956).

the east (74°07'N, 23°30'W) to the nunataks of Adolf Hoel Gletscher (74°03'N, 26°45'W) in the west. In the north, most of the localities are grouped around c. 74°20' N with the exception of one locality at 74°27'N, 24°09'W.

Within WA4 there are three regions with a different physiography. Promenadedal is a dry valley with large terraces at altitudes between 250 m and 300 m (Fig. 26). They form a kind of "stair" along the deep gorge of the river. Northeast of Vibeke Sø a steep slope leads to a ridge of 800-900 m at the edge of a plateau. The level of the lake may change considerably, as it has been seen to do during several observations in the summer 1956. The region around the basecamp at Krumme Langsø (74°06'N, 24°26'W, ca. 200 m) lies in the flat and open valley leading to the Sødal. Several terraces with a few ponds connect the lake with the dry hills and the plateau separating Krumme Langsø from Vibeke Sø. The third region includes the nunataks north of 74°N at altitudes between 450 m and 1400 m.

The two base-camps at Vibeke Sø (1952) and at Krumme Langsø were established by sea-planes of the Danish East Greenland Expeditions. The nunataks were reached by helicopter stationed at Mesters Vig and chartered

Fig. 26. Aerial view of Promenadedal (WA4, 1956).

by Dr. L. Koch for five days (August, 14.-18th 1956). Details of the flight are given in Schwarzenbach (1961).

Geology

H. R. Katz, W. Diehl and H. Röthlisberger were transported by weasels of the Expéditions Polaires Françaises (P.-E.Victor) from Cecilia Nunatak to the nunataks of Hobbs Land (73°57'N). They crossed the nunataks via Gerard de Geer Gletscher to Nordfjord (Katz 1952, 1953).

Haller (1956 a) has described and mapped the geology of the study area. Further maps have been published by Haller (1956 b, 1970, 1983), Koch & Haller (1971). Haller has observed that a rather broad belt of carbonatic bedrocks cuts the area in direction from NE to SW. This belt separates the sediments (poor in carbonates) of Hudson Land in the southeast from the crystalline of the nunataks and of the Nørlund Alper.

Erratic glacial deposits have been found at six places in the nunataks. This suggests, that at least some nunataks have been covered by ice at one time.

Climate

Hovmøller (1947) has summarized the weather observations 1922/37 of the Norvegian station at Myggbukta, Hold with Hope (73°29'N, 21°34'W). This station shows, that the conditions of the outer coast have rather low mean temperatures in the summer (average monthly temperatures: June +1.4°C, July +3.9°C, August +3.1°C). The mean annual precipitation was 220 mm. According to Ohmura and Reeh (1991) an annual total precipitation of about 200 mm might be expected. Bay (1992) has summarized the data from Daneborg for the period 1958-1987 (Tab. 2).

The climate of the area WA4 is much more continental than that recorded at Myggbukta and Daneborg stations. During the summer the temperature can be about 2-3°C higher and precipitation is lower. There are also less days with clouds, and fog seems to be rare. The area is protected by the mountains of Hudson Land against the humid wind from the east. Dry and often strong

Table 2. Mean temperatures, precipitation and 'degree-days above 0°C' (1958/87). Daneborg (74°18'N, 20°13'W) cited from Bay (1992)

	A	B	C	D	E
mean	– 9.9°	3.6°	285.5 mm	45.7 mm	285.2
min.	–11.3°	2.5°	129.2 mm	14.7 mm	205.1
max.	– 9.3°	4.6°	477.8 mm	128.9 mm	376.9

A mean annual temperature °C
B mean summer temperature (01/07-31.08) °C
C mean annual precipitation in mm
D mean summer precipitation (01/07-31/08) in mm
E degree-days above 0°C

föhn winds blow with higher frequencies and higher intensity in the western nunataks causing early snow melt. Salt-crusts on the surface of the soil are a result of these dry winds, and this has been observed around the base camp at Krumme Langsø.

2.5.2 Botanical exploration and vegetation

Botanical exploration

Gelting (1934) mentions the first collections from the area:

- Jordanhill (ca. 74°07'N,22°45'W): G. Seidenfaden, 25/07/1930.
- Promenadedal (ca. 74°06'N, 22°45'W): G. Seidenfaden, 25/07/1930 and T. Johansen in the winter 1932/33.
- Scotstounhill (ca. 74°15'N, 22°34'W, 200 m): K. Teichert.
- C. H. Ostenfeld Nunatak (ca. 74°20'N, 22°47'W, 600 m): K. Teichert.
- Keglebjerg (ca. 74°33'N, 23°18'W, 100-200 m above the level of the inland ice): T. Johansen, March 1931.
- Marianne Nunatak (ca. 74°07'N, 23°25'W, 1600 m): A. Schwarck, April 1932.
- Mt. Eremit (74°15'N, 23°26'W, 1800 m): A. Schwarck, April 1932, collection of *Poa abbreviata.*

H. Röthlisberger has made botanical observations and has collected plants during his journey with H. R. Katz and W. Diehl from the westernmost nunataks of Hobbs Land to Nordfjord in August, 1951. These observations and the lists of the collected species are published by Schwarzenbach (1961), who had the chance to see the specimens which were identified by Walo Koch and kept at the herbarium of the ETH Zürich (now transfered to the herbarium of the University of Zürich, Switzerland).

The results of the botanical investigation in WA4 are published by Schwarzenbach (1961) based on the identifications by K. Holmen, Th. Sørensen and G. Halliday. C. Bay has revised the material and included it in his paper showing dot-maps for the distribution of all species found in Greenland north of 74°N (Bay 1992). Halliday (pers. comm.) intends to include the material in his publication on the phytogeography of East Greenland.

Vegetation

The number of species (126) is surprisingly high and is explained by the great variety of ecological conditions within the area (Schwarzenbach 1961). A few rather rare species, such as *Potamogeton filiformis, Hippuris vulgaris, Ranunculus confervoides, Rumex acetosella, Carex rariflora,* and *C. microglochin* have been found. *Ranunculus glacialis* was common at several places on the nunataks up to 1370 m.

Some species have been found at unusually high altitudes: *Campanula uniflora* (1470 m), *Hierochloë alpina* (1270 m), *Melandrium apetalum* (1400 m), *Dryas octopetala s.l.* (930 m), *D. integrifolia s.str.* (930 m), *Alopecurus alpinus* (1200 m), and *Poa alpina s.l.* (900 m).

An outpost of the *Betula*-heath with *B. nana, Pedicularis lapponica, Tofieldia pusilla* and *Poa pratensis s.l.* has been found at 350 m east of Vibeke Sø. *Betula nana* was also present east of Krumme Langsø (260 m) together with *Arctostaphylos alpina,* and *Pedicularis flammea.* An example of the typical *Vaccinium*-heath has been observed in Ole Rømer Land, northeast of Waltershausen Gletscher (74°08'N, 24°26'W, 450 m). This type of vegetation is otherwise only present in fragments west of Vibeke Sø and near the base camp at Krumme Langsø. *Cassiope tetragona* has an isolated occurence in Bartholin Land (74°14'N, 25°15'W, 910 m). *Rhododendron lapponicum* was found up to 750 m in Promenadedal.

Calcicolous species and even halophytes have been found at dry places in the zone of crystalline rocks. This occurence is explained by the high content of mineral salts at the surface of the soil caused by the evaporation of soil-moisture by dry föhn winds during the summer. The typical species of acidic soils do not tolerate these extreme conditions while calciphilous plants and halophytes can still grow in places with a considerable concentration of ions in the soil (Schwarzenbach 1960). The emigration of typical calciphytes to sites on crystalline substratum has been found by the author in other continental areas of East Greenland too (e.g. Hjørnedal, Gåse Land, 70°N, 1994) also in North Greenland (North Peary Land, Grønnemark, 83°N, 1995). One of the best examples of this type of vegetation has been observed in Ole Rømer Land, northeast of Waltershausen Gletscher (74°08'N, 24°26'W). This include *Armeria scabra* ssp. *sibirica, Arnica angustifolia, Rumex acetosella s.l., Festuca vivipara, Calamagrostis purpurascens, Carex rupestris, C. supina* ssp. *spaniocarpa, Dryas octopetala s.l., Hierochloë alpina, Kobresia myosuroides,* and *Lesquerella arctica.*

Snow-patches are present on the nunataks. On one of these in West Bartholin Land (74°21'N, 25°58'W, 1250 m) *Alopecurus alpinus, Phippsia algida s.l., Luzula confusa, Cardamine bellidifolia, Saxifraga nivalis,* and *Saxifraga cernua* have been collected. At another snow-patch 23 species have been observed above Vibeke Sø (780 m). These include *Ranunculus glacialis, R. sulphureus, Saxifraga platysepala, S. hyperborea, S. nathorstii, Eutrema edwardsii, Cardamine bellidifolia, Koenigia islandica* (Fig. 27), *Sagina intermedia, Taraxacum arcticum, Colpodium vahlianum, Arctagrostis latifolia,* and *Alopecurus alpinus.*

In Bartholin Land (74°19'N, 25°25'W, 1220 m) there was a habitat with *Carex rupestris* as the dominant species, and also with *C. misandra, C. nardina, Kobresia myosuroides, Festuca baffinensis, Poa arctica, P. glauca, Trisetum spicatum, Potentilla hookeriana s.l., P. hyparctica, Cerastium arcticum s.l., Minuartia rubella, Campanula uniflora,* and *Taraxacum phymatocarpum.*

Fig. 27. Koenigia islandica. *Vibeke Sø (WA4, 780 m, 1952).*

The fell-field vegetation shows a clear difference depending on the kind of bedrock. Calcareous and crystalline rocks are found in Bernhard Studer Land (74°03'N, 26°45'W, 1170-1210 m). *Draba bellii, D. arctica s.l., Lesquerella arctica, Braya purpurascens, Carex nardina* and *Trisetum spicatum* have been found on the calcareous soil. *Saxifraga caespitosa s.l., S. nivalis, S. cernua, Erigeron eriocephalus, Poa abbreviata,* and *Festuca brachyphylla* have been collected on the crystalline rocks.

2.6 Kronprins Christian Land, Ingolf Fjord, Hekla Sund: Study area WA5 (Fig. 28)

80°04'-80°39'N; 19°53'-24°10'W; altitude range: 0-1410 m
There are four sub-areas which have been visited:

- Centrum Sø, Drømmebjerg: 31.07.-09.08.1952 and 17.08.-22.08.1952. E. Fränkl, F. Müller, F. H. Schwarzenbach
- Månevig, plateau N of Ingolf Fjord: 10.08.-12.08.1952. E. Fränkl, F. H. Schwarzenbach

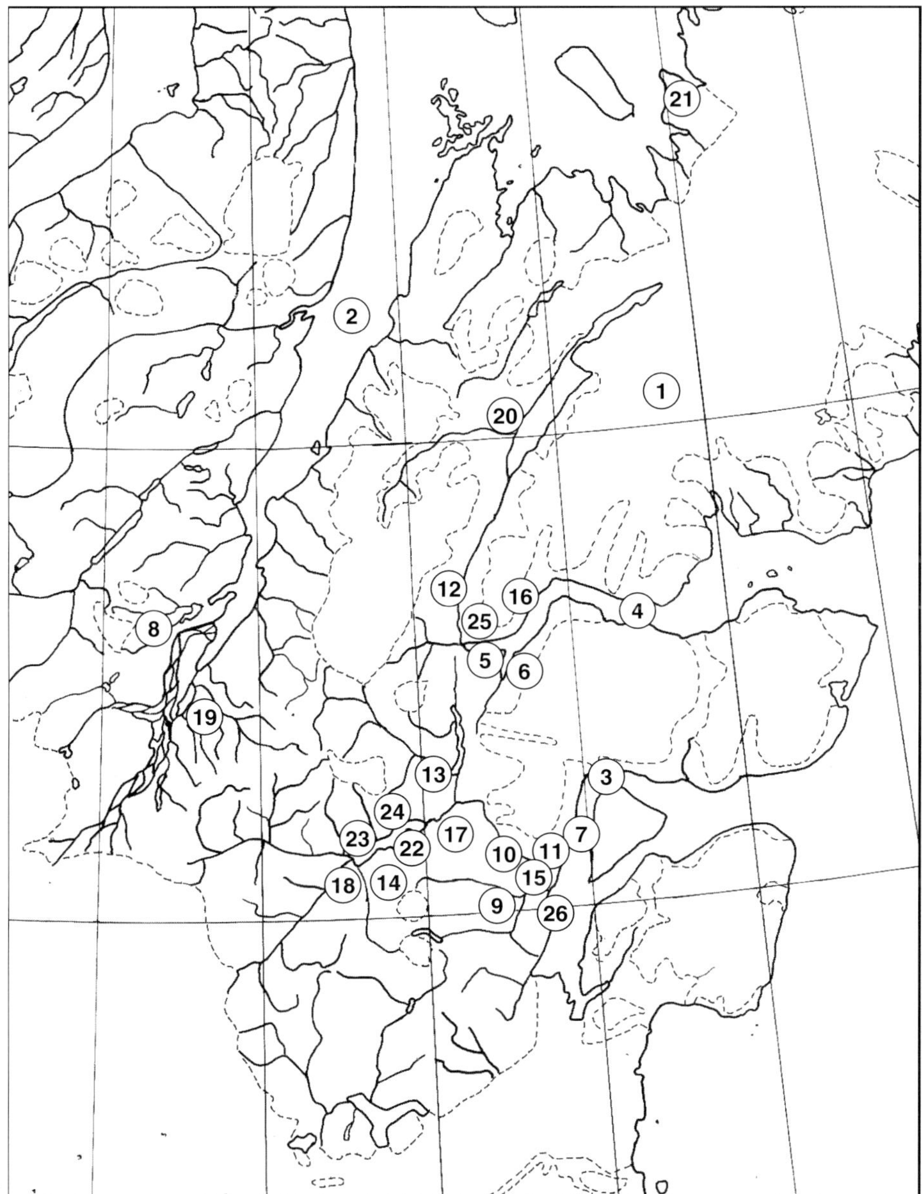

Fig. 28. Map of WA5, Kronprins Christian Land, Ingolf Fjord, Hekla Sund: (1) Kronprins Christians Land, (2) Danmark Fjord, (3) Hekla Sund, (4) Ingolf Fjord, (5) Månevig, (6) Solvig, (7) Marmorvigen, (8) Campanuladal, (9) Rivieradal, (10) Sæfaxi Elv, (11) Holbæk Bugt, (12) Vandredalen, (13) Drømmebjerg, (14) Flyverbjerg, (15) Posten, (16) Tågefjeldene, (17) Ulvebjerg, (18) Centrum Sø, (19) Fyn Sø, (20) Romer Sø, (21) Station Nord, (22) Base camp, (23, 24) Advanced camps north of Centrum Sø, (25) Camp at Ingolf Fjord, (26) Camp at Hekla Sund.

- Mountains west of Hekla Sund: 13.08-15.08.1952. E. Fränkl, F. H. Schwarzenbach
- Danmark Fjord, Fyn Sø, 16.08.1952. E. Fränkl, F. H. Schwarzenbach

2.6.1 Landscape, geology, climate

Landscape

The names of the area are based on Higgins (1994). The study area can be described as the belt between 80°03′-80°38′N reaching from the Head of Centrum Sø in the west and to the innermost part of Ingolf Fjord and to Hekla Sund in the east. A one-day flight to Danmark Fjord (Holbæk Bugt 80°39′N, 23°35′W) and Fyn Sø (80°31′N, 24°10′W) extended the area to NW.

Centrum Sø (80°10′N, 22°00′W) is more than 25 km long and up to 4 km broad (Figs. 29, 30). The lake is situated in a flat valley running from west to east. The Sæfaxi Elv flows from the lake to Marmorvigen, Hekla Sund (80°05′N, 20°08′W). A low pass leads from the southern shore of Centrum Sø through the Rivieradal to the coast of Hekla Sund/Djimphna Sund. The Rivieradal separates the mountains around Flyverbjerg (80°08′N, 21°52′W, >1000 m) in the east and Ulvebjerg (80°09′N, 21°31′W, >1000 m) in the west. In the northernmost part of Rivieradal there are very large polygons (± 60 m diameter separated by deep spaces ± 1 m in wide). North of Centrum Sø there are the mountains and plateaus of the Drømmebjerg exceeding the altitude of 1100 m. Moving-soils and active polygon-fields are common at the slopes and on the plateaus.

Fig. 29. Arrival of the Danish Catalina aircraft at Centrum Sø (WA5, Kronprins Christian Land, 22.08.1952).

Fig. 30. Evening in Kronprins Christian Land (WA5, 1952).

Conspicuous terraces (Fig. 32) mark the northern coast of the innermost Ingolf Fjord (Månevig 80°32'N, 20°29'W). At the eastern end of the fjord is the broad valley Vandredalen. It runs from Marmorvigen in the south to Romer Sø in the north. The mountain chain of Tågefjeldene (80°38'N, 19°53'W) passes the altitude of 1100 m.

Posten (80°03'N., 20°12'W) is a mountain south of Marmorvigen where the Sæfaxi Elv reaches Hekla Sund.

Geology

The results of the geological work of summer 1952 have been published by Adams & Cowie (1953) and Fränkl (1954, 1955). Since then, Grønlands Geologiske Undersøgelse (GGU) has investigated the geology of North Greenland. A good introduction is given by Ghisler (1986) and by Peel & Sønderholm (1991) with a geological map of North Greenland 1:1000000 compiled by H.-J. Bengård & N. Henriksen (1991). The Upper Proterozoic Rivieradal Sandstones form a broad belt running from south to north from 79°N to Romer Sø (ca. 80°30') which is cut off by a major thrust in the east passing the Centrum Sø. North of the lake there is a puzzling mosaique of Ordovician and Siluarian sediments along Vandredalen and the chain of Drømmebjerg (Haller 1983).

Fig. 31. Centrum Sø (WA5, 1952).

Fig. 32. Aerial view of terraces at the northern side of Ingolf Fjord (WA5, 1952).

Climate

The nearest weather-stations are Danmarkshavn (76°46'N, 18°46'W) and Station Nord (81°36'N, 16°40'W). Tab. 3 shows the mean temperatures and precipitation for the whole year and the summer (01/07-31/08) as well as the number of "deegree-days above 0°C" cited from Bay (1992, p. 6).

The means of temperatures and precipitations of WA5 may be intermediate between the values from Danmarkshavn and Station Nord, if the situation of the outer coast (Ingolf Fjord, Hekla) is considered. The climate of Centrum Sø seems to be more continental. The influence of temperature inversions and fog along the outer coast can be recognised by the vertical distribution of vascular plants: A remarkably high number of species (10) are growing at higher levels between 800 m and 900 above the clouds of the inversion-layer. There were many snow-patches and snow-beds on the mountains and plateaus west of Drømmebjerg as well as on the Tågefeldene.

2.6.2 Botanical exploration and vegetation

Botanical exploration

The history of the botanical exploration of North Greenland was summarized by Bay (1992). The first botanical colllection in WA5 was undertaken by Schwarzenbach in 1952 (Schwarzenbach 1954). Bay and Fredskild (1994) have visited four localities in the coastal area of Holm Land and Amdrup Land between 80°20'- 80°57'N and 14°40'-18°45'W, and a place on the island Henrik Krøyer Holme 80°42'N, 13°50'W.

Table 3. Mean temperatures, precipitation and 'degree-days above 0°C' (1961/87) cited from Bay (1992). Danmarkshavn (76°46'N, 18°46'W) and Station Nord (81°36'N, 16°40'W)

	A	B	C	D	E
Danmarkshavn					
mean	–12.4°	3.0°	192.2 mm	44.0 mm	236.7
min.	–13.6°	2.3°	74.4 mm	7.0 mm	171.5
max.	–10.4°	4.1°	397.2 mm	118.4 mm	346.1
Station Nord					
mean	–17.1°	2.3°	227.3 mm	48.1 mm	221.6
min.	–18.4°	0.1°	98.2 mm	15.7 mm	119.6
max.	–15.5°	xxx	422.7 mm	116.7 mm	338.8

A mean annual temperature °C
B mean summer temperature (01/07-31.08) °C
C mean annual precipitation in mm
D mean summer precipitation (01/07-31/08) in mm
E degree-days above 0°C

Vegetation

A total of 93 species of vascular plants have been collected. *Campanula uniflora* has been found at Danmark Fjord (80°39'N, 23°35'W). *Carex atrofusca* and *C. supina* ssp. *spaniocarpa* (northernmost record 80°13'N) were growing near Centrum Sø. *Pedicularis flammea* has been collected in the Sæfaxi Dal (80°13'N, 20°48'W). *Vaccinium uliginosum* ssp. *microphyllum* has been observed twice, but unfortunately the collected specimen was lost. In the westernmost part of WA5 (Fyn Sø) the heath was represented by *Dryas integrifolia s.str.* and *Cassiope tetragona* with *Carex misandra*. *Eriophorum callitrix* has been found in a mire. Near the shore *Braya purpurascens, B. thorild-wulffii, Cerastium regelii, Colpodium vahlianum, Minuartia rossii,* and *Kobresia simpliciuscula* were found.

Carex atrofusca, C. supina ssp. *spaniocarpa, Braya thorild-wulffii, Epilobium arcticum* and *Elymus hyperarcticum* have been found along Centrum Sø. A type of dry heath with *Calamagrostis purpurascens, Poa glauca, Trisetum spicatum, Kobresia myosuroides,* and *Pedicularis flammea* was observed in the Sæfaxi Dal. *Arctagrostis latifolia, Cerastium regelii, Cochlearia groenlandica, Colpodium vahlianum, Draba bellii, Eutrema edwardsii, Minuartia rossii, Ranunculus sulphureus, Sagina intermedia* and *Stellaria longipes* s.l have been collected in the Rivieradal. Snow-bed and snow-patch vegetation were found at many places on the plateau and the mountains around Drømmebjerg: e.g. *Alopecurus alpinus, Braya thorild-wulffii, Cardamine bellidifolia, Carex glacialis* (670 m), *Cerastium regelii* (955 m), *Draba adamsii s.l., D. alpina s.l., D. lactea, Epilobium arcticum, Minuartia rossii, Phippsia algida s.l., Puccinellia angustata* (810 m), *P. bruggemanni* (600 m), *Saxifraga platysepala* (650 m), and *Taraxacum arcticum* (670 m).

Only 48 species have been collected at the coast of Ingolf Fjord (Månevigen) and on the Tågefjeldene at altitudes up to 470 m.

Alopecurus alpinus, Cerastium regelii, Cochlearia groenlandica, Deschampsia brevifolia, Draba arctica s.l., D. nivalis, Equisetum variegatum, Eutrema edwardsii, Festuca baffinensis, Lesquerella arctica, Phippsia algida s.l., Ranunculus affinis, Saxifraga platysepala, and *Trisetum spicatum* have been found on the terraces below 70 m. *Calamagrostis purpurascens* (330 m), *Carex nardina* (410 m), *C. rupestris* (330 m), *Cassiope tetragona* (470 m), *Cystopteris fragilis* (330 m), *Draba bellii* (470 m), *D. lactea* (330 m), *D. subcapitata* (470 m), *Festuca hyperborea* (470 m), *Hierochloë alpina* (140 m), *Melandrium affine s.l.* (330 m), *M. triflorum s.str.* (410 m), *Minuartia rossii* (150 m), *M. rubella* (330 m), *Potentilla rubricaulis* (470 m), *Puccinellia angustata* (330 m), and *Taraxacum arctogenum* (470) were growing along a steep south-facing slope.

3. Taxonomic considerations

3.1 Introduction

Generally, the taxonomy follows Böcher et al. (1978), Bay (1992) and Fredskild (1996 b). Exceptions will be discussed in chapter 3.3.

The botanical observations in the nunatak area north of 74°N have been published by Schwarzenbach (1961). The area is defined in the same way in this study and is now called study area WA4. The collected specimens of 1956 have been identified by Holmen and Sørensen according to the state of knowledge of that time. Since then Bay (1992) has revised the herbarium specimens and included them in his publication. The old and the revised names are listed in Tab. 4.

3.2 List of species

Each species is listed and classified to the proposed "**A**ltitude **D**istribution **T**ype, ADT" (Tab. 5) and the **N**orth **G**reenland **D**istribution **T**ype, NGDT (Bay

Table 4. Changes of nomenclature since Schwarzenbach (1961)

Old names (used 1961)	Revised names
Braya intermedia Th. Sør.	*B. humilis* (C. A. Mey.) Robins. (see 3.3)
Campanula rotundifolia L. coll.	*C. gieseckiana* Vest in R. & S. *s.l.*
Carex amblyorhyncha Krezc. ssp. *pseudolagopina* Th.Sør.	*C. marina* Dew. *s.l.* (3.3)
Cerastium alpinum L. coll.	*C. arcticum* Lge. *s.l.* (3.3)
Draba arctogena Ekm./ *D. groenlandica* Ekm.	*D. arctogena* Ekm.
Draba cinerea Adams	*D. arctica* J. Vahl *s.l.* (3.3)
Draba hirta L.	*D. glabella* Pursh
Draba oblongata R.Br. incl. *D. micropetala* Hook.	*D. adamsii s.l.* (3.3)
Dryas chamissonis Spreng., *D. cf. crenulata* Juz., *D. octopetala* L. ssp. *punctata* Juz.	*D. octopetala s.l.* (3.3)
Poa pratensis L. coll., *P. alpigena* (E. Fr.) Lindm.	*P. pratensis* L. *s.l.*
Ranunculus pedatifidus Sm. ssp. *affinis* (R. Br.) Hultén	*R. affinis* R. Br. *s.l.*
Saxifraga caespitosa L. ssp. *caespitosa* var. *uniflora* R. Br.	*S. caespitosa* L. *s.l.*
Saxifraga flagellaris Willd. ssp. *platysepala* (Trautv.) A. E. Porsild	*S. platysepala* (Trautv.) Tolm.
Saxifraga hyperborea R. Br.	*S. rivularis* L. *s.l.* (3.3)
Stellaria ciliatosepala Trautv.	*S. longipes* Goldie *s.l.* (3.3)
Stellaria crassipes Hultén	*S. longipes* Goldie *s.l.* (3.3)

Table 5. List of species collected in WA1-WA5

Species	ADT	NGDT	C
Alopecurus alpinus Sm.	D2	1	–
Antennaria canescens (Lge.) Malte	X	12	–
Antennaria porsildii Ekm.	X	14	–
Arabis alpina L.	G3	14	–
Arabis holboellii Horn.	G3	*	–
Arctagrostis latifolia (R. Br.) Griseb.	D2	2a	–
Arctostaphylos alpina (L.) Spreng.	B3	14	–
Arenaria humifusa Wbg.	G3	9	–
Arenaria pseudofrigida (Ostf. & Dahl) Juz.	X	7a	–
Armeria scabra Pall. ssp. *sibirica* (Turcz.) Hyl.	X	8a	–
Arnica angustifolia M. Vahl in Hornem.	B2	5a	–
Betula nana L.	B3	12	–
Botrychium lunaria (L.) Sw.	G3	*	–
Braya humilis (C. A. Mey.) Robins.	A2	7a	C
Braya linearis Rouy	X	14	–
Braya purpurascens (R. Br.) Bge.	A2	2a	–
Braya thorild-wulffii Ostf.	H1	4	–
Calamagrostis purpurascens R. Br.	A2	8a	–
Campanula gieseckiana Vest in R. & S. *s.l.*	B3	12	–
Campanula uniflora L.	B1	5a	–
Cardamine bellidifolia L.	C1	1	–
Carex atrofusca Schkuhr	D2	8a	–
Carex bicolor All.	G3	*	–
Carex bigelowii Torr. *s.l.*	X	5	C
Carex capillaris L. *s.l.*	B2	5a	C
Carex glacialis Mack.	X	6c	–
Carex glareosa Wbg.	F3	6b	–
Carex lachenalii Schkuhr	G3	6b	–
Carex marina Dew. *s.l.*	G3	.8a	C
Carex maritima Gunn.	C2	2b	–
Carex microglochin Wbg.	G3	14	–
Carex misandra R. Br.	B1	1	–
Carex nardina Fr.	A1	1	–
Carex norvegica Retz.	X	14	–
Carex parallela (Laest.) Sommerf.	G3	14	–
Carex rariflora (Wbg.) Sm.	G3	14	–
Carex rupestris All.	A2	8a	–
Carex saxatilis L.	D3	8a	–
Carex scirpoidea Michx.	B3	12	–
Carex stans Drej. *s str.*	H2	1c	C
Carex supina Wbg. ssp. *spaniocarpa* (Steud.) Hult.	A2	5	–
Carex ursina Dew.	F3	5a	–
Cassiope tetragona (L.) D. Don	B1	1	–
Cerastium arcticum Lge. *s.l.*	B1	1	C
Cerastium regelii Ostf. ssp. *caespitosum* (Malmgr.) Tolm.	H1	3	–
Chamaenerion latifolium (L.) Sweet	B2	1c	–
Cochlearia groenlandica L.	X	1	–
Colpodium vahlianum (Liebm.) Nevski	C2	2	–
Cystopteris fragilis (L.) Bernh. *s.l.*	B1	1c	C
Deschampsia brevifolia R. Br.	H2	2b	–

Species	ADT	NGDT	C
Draba adamsii Led. *s.l.*	X	2	C
Draba alpina L. *s.l.*	X	8a	C
Draba arctica J. Vahl *s.l.*	B1	2b	C
Draba arctogena Ekm.	X	2a	–
Draba aurea M. Vahl	G3	*	–
Draba bellii Holm	A1	2	C
Draba crassifolia Grah.	G3	14	–
Draba fladnizensis Wulf.	A2	7b	C
Draba glabella Pursh	B2	5	–
Draba gredinii Ekm.	X	*	C
Draba lactea Adams	B1	1	C
Draba nivalis Liljebl.	A1	5	–
Draba subcapitata Simm.	B1	2	–
Dryas integrifolia M. Vahl *s.str.*	A2	1a	C
Dryas octopetala L. *s.l.*	A2	7	C
Elymus hyperarcticus (Polun.) Tzvel.	H2	7a	–
Empetrum nigrum L. ssp. *hermaphroditum* (Hagerup) Boech.	X	5	–
Epilobium arcticum Sam.	H1	7a	–
Equisetum arvense L. *s.l.*	C3	1c	C
Equisetum variegatum Schleich.	C3	2b	–
Erigeron compositus Pursh	X	2b	–
Erigeron eriocephalus J. Vahl	B1	8a	–
Erigeron humilis Grah.	C3	7	–
Eriophorum callitrix Cham.	D3	8a	–
Eriophorum scheuchzeri Hoppe	D3	1b	–
Eriophorum triste (Th.Fr.) Hadac & Löve	D2	2	–
Euphrasia frigida Pugsl.	B3	7	–
Eutrema edwardsii R. Br.	C3	2b	–
Festuca baffinensis Polun.	B1	2b	–
Festuca brachyphylla Schult. & Schult.	B1	1c	–
Festuca hyperborea Holmen ex Frederiksen	X	1	–
Festuca rubra L. *s.l.*	B3	14	–
Festuca vivipara (L.) Sm.	X	7a	–
Gentiana detonsa Rottb.	G3	*	–
Gentiana tenella Rottb.	G3	7	–
Harrimanella hypnoides (L.) Coville	G3	12	–
Hierochloë alpina (Willd.) R. & S.	A2	1	–
Hippuris vulgaris L.	E3	8a	C
Honckenya peploides (L.) Ehrh. var. *diffusa* (Horn.) Mattf.	F3	8	–
Huperzia selago (L.) Bernh. ex Schrank & Mart. ssp. *arctica* (Grossh.) Löve & Löve	X	5a	–
Juncus arcticus Willd.	C3	14	–
Juncus biglumis L.	D1	1	–
Juncus castaneus Sm.	D3	5a	–
Juncus trifidus L.	G3	14	–
Juncus triglumis L.	D2	2b	–
Kobresia myosuroides (Vill.) Fiori & Paol.	A2	1c	–
Kobresia simpliciuscula (Wbg.) Mack.	D2	8a	–
Koenigia islandica L.	C3	5a	–
Lesquerella arctica (Wormsk.) S. Wats.	A2	1c	–

Species	ADT	NGDT	C
Luzula arctica Blytt	C1	1	–
Luzula confusa Lindeb.	C1	1	–
Luzula spicata (L.) DC.	X	7	–
Lycopodium annotinum L. ssp. *alpestre* (Hartm.) Löve & Löve	G3	*	–
Melandrium affine J. Vahl *s.l.*	B1	1c	C
Melandrium apetalum (L.) Fenzl ssp. *arcticum* (Fr.) Hult.	C1	2	–
Melandrium triflorum (R. Br.) J. Vahl *s.str.*	B2	1b	C
Minuartia biflora (L.) Sch. & Th.	C2	5	–
Minuartia rossii (R. Br.) Graebn.	X	2a	–
Minuartia rubella (Wbg.) Hiern	B1	1	–
Minuartia stricta (Sw.) Hiern	G3	8	–
Oxyria digyna (L.) Hill	C2	1	–
Papaver radicatum Rottb. *s.l.*	B1	1	C
Pedicularis flammea L.	B2	5	–
Pedicularis hirsuta L.	B2	1	–
Pedicularis lapponica L.	B3	14	–
Phippsia algida (Sol.) R. Br. *s.l.*	X	*	C
Pinguicula vulgaris L.	G3	*	–
Pleuropogon sabinei R. Br.	H2	2b	–
Poa abbreviata R. Br.	A1	2	–
Poa alpina L. *s.l.*	X	14	C
Poa arctica R. Br.	B1	1	–
Poa glauca M. Vahl	A1	1b	–
Poa hartzii Gandoger	B2	2b	–
Poa pratensis L. *s.l.*	X	6b	–
Polygonum viviparum L.	B2	1	–
Potamogeton filiformis Pers.	E3	7	–
Potentilla crantzii (Cr.) G. Beck	G3	14	–
Potentilla hookeriana Lehm. *s.l.*	A1	1C	C
Potentilla hyparctica Malte	X	1	–
Potentilla nivea L. emend. Hult.	B1	5	C
Potentilla rubricaulis Lehm.	B2	2b	–
Puccinellia angustata (R. Br.) Rand. & Redf.	H1	2	–
Puccinellia bruggemanni Th. Sør.	H2	3a	C
Pyrola grandiflora Rad.	X	6	–
Ranunculus affinis R. Br. *s.l.*	C2	8a	–
Ranunculus confervoides Fr.	E3	8a	–
Ranunculus glacialis L.	X	7	–
Ranunculus hyperboreus Rottb.	C2	1b	–
Ranunculus nivalis L.	G3	8a	–
Ranunculus pygmaeus Wbg.	X	5	–
Ranunculus sulphureus Sol. in Phipps	C2	1	–
Rhodiola rosea L.	G3	14	–
Rhododendron lapponicum (L.) Wbg.	B3	5	–
Rumex acetosella L. *s.l.*	B3	14	–
Sagina caespitosa (J. Vahl) Lge.	G3	5	–
Sagina intermedia Fenzl	X	1	–
Salix arctica Pall.	B1	1	–
Salix herbacea L.	X	5	–
Saxifraga aizoides L.	D2	8a	–

Species	ADT	NGDT	C
Saxifraga caespitosa L. *s.l.*	B1	1	–
Saxifraga cernua L.	B1	1	–
Saxifraga foliolosa R. Br.	D3	1	–
Saxifraga hieracifolia W. & K.	X	7a	–
Saxifraga nathorstii (Dusén) Hayek	D2	14	–
Saxifraga nivalis L.	B1	1	C
Saxifraga oppositifolia L.	A1	1	–
Saxifraga platysepala (Trautv.) Tolm.	H2	2a	–
Saxifraga rivularis L. *s.l.*	X	5	C
Saxifraga tenuis (Wbg.) H. Smith	C1	1	C
Sibbaldia procumbens L.	G3	14	–
Silene acaulis (L.) Jacq.	B2	1b	–
Stellaria humifusa Rottb.	F3	5	–
Stellaria longipes Goldie *s.l.*	H2	1	C
Taraxacum arcticum (Trautv.) Dahlst.	X	3a	–
Taraxacum arctogenum Dahlst.	H2	4b	–
Taraxacum brachyceras Dahlst.	G3	14	–
Taraxacum phymatocarpum J. Vahl	A1	1	–
Taraxacum pumilum Dahlst.	H2	3a	–
Thalictrum alpinum L.	G3	14	–
Tofieldia coccinea Richards.	B3	8	–
Tofieldia pusilla (Michx.) Pers.	B3	12	–
Triglochin palustre L.	G3	14	–
Trisetum spicatum (L.) Richt.	B1	1c	–
Vaccinium uliginosum L. ssp. *microphyllum* Lge.	B2	5a	–
Viscaria alpina (L.) G. Don	G3	*	–
Woodsia glabella R. Br.	B2	14	–

Abbreviations:
ADT Altitude Distribution Type
NGDT North Greenland Distribution Type (Bay 1992)
C Taxonomic comments in section 3.3.

1992). Column "C" indicates that the species is discussed in the following section 3.3 under "Taxonomic comments".

3.3 Taxonomic comments

Braya humilis (C. A. Mey.) Robins. incl. *B. intermedia* Th. Sør.
Böcher et al. (1978:119) treat *B. intermedia* as a species. One argument for this taxonomical separation might be that *B. intermedia* has a number of chromosomes (2n = 70) differing from *B. linearis* (2n = 42) and *B. humilis* (2n – 42). On the contrary, Bay (1992) includes it in *B. humilis.*

Th. Sørensen has determined specimens collected in WA4, Krumme Langsø as *B. intermedia*. The plants were growing together with *B. purpurascens, B. humilis* and *B. linearis*. Based on the revision by C. Bay these specimens are included in *B. humilis*.

Carex bigelowii Torr. *s.l.* / *C. stans* Drej. *s.str.*
According to Fredskild (1996 a) the distinction between *C. bigelowii* and *C. stans* causes difficulties where the two species co-exist in the same area. Hybrids may occur. The few specimens of doubtful hybrid origin are excluded from the present study. Plants from Kronprins Christian Land (WA5) matching the identification of *C. stans* are classified as *C. stans* Drej. *s.str.* All other plants are included under the highly variable species *C. bigelowii* Torr. *s.l.*

Carex capillaris L. *s.l.*
According to Böcher et al. (1978: 270) *C. capillaris* L. *s.l.* forms a group of several small species, three of them are found in Greenland: *C. capillaris* ssp. *fuscidula* (Krecz.) Löve & Löve, *C. boecheriana* Löve, Löve & Raymond (syn. *C. capillaris* var. *robustior* Drej. ex Lange), and *C. krausei* Böckeler ssp. *porsildiana* (Polun.) Löve, Löve & Raymond.

According to Fredskild (1996 a) only ssp. *fuscidula* (Krecz.) Löve & Löve and ssp. *robustior* Drej. ex Lange occur in Northeast Greenland, the latter only in the continental inland.

All plants found in WA1-WA5 are classified as *C. capillaris* L. *s.l.*

Carex marina Dew. *s.l.*
Böcher et al. (1978: 257) distinguish between ssp. *marina* Dew. and ssp. *pseudolagopina* (Th. Sør.) Böcher. According to Halliday (pers. comm.) the correct name is *Carex marina* ssp. *pseudolagopina* (Böcher) Böcher. Following Fredskild (1996 a) the two subspecies are united as *C. marina* Dew *s.l.*

Carex stans Drej. *s.str.* (see *C. bigelowii* Torr. *s.l.*)

Cerastium arcticum s.l. Lange, *C. alpinum* L. ssp. *lanatum* (Lam.) Hegetschw.
The taxonomy of the highly polymorphous plants *C. alpinum s.l.* and *C. arcticum s.l.* is often discussed (e.g. Murbeck 1898, Seidenfaden 1933, Gelting 1934, Holmen 1957, Fredskild 1966, Scoggan 1978, Feilberg 1984, Bay 1992). This is the case, too, for related taxa in the Alps. Thus, Landolt (1964) states that in the Swiss Alps eight species – difficult to distinguish – are found above 1500 m. Therefore, it might be of some interest to summarize the history of this long-lasting controversial discussion.

Linné has given a very brief diagnosis of *C. alpinum* (cited from Gelting 1934: 35): "*foliis ovato-lanceolatis caule diviso, capsulis oblongis.*" This short description (based probably on Scandinavian plants) which includes the key characteristics used by alpine or arctic floras as Schinz & Keller (1923), Binz & Becherer (1976), Böcher et al. (1978) has caused much incertainty.

Hegetschweiler (1825) basing on observations in the eastern part of the Swiss Alps has proposed a variety cited by Gelting (1934) as var. *lanatum* (Lam.) Hegetschw. (not var. *lanatum* [(Lam.) Asch. & Graebn.)]. Later the alpine *Cerastia* have been split into several ecologically and/or morphologically different species which are difficult to distinguish. According to Schinz & Keller (1923) the following species are found in the Swiss Alps: *C. latifolium* L., *C. uniflorum* Clairv., *C. pedunculatum* Gaudin, *C. alpinum* L., *C. arvense* (L.) Thuill. and *C. caespitosum* Gilib. Concerning *C. alpinum* Schinz & Keller (1923) remark: "often confused". Hess et al. (1967) state: "*C. alpinum* bildet mit 8 anderen Arten eine arktisch-alpine Gruppe (vgl. Hultén 1956). Im Gebiet (der Schweiz) treten gelegentlich verkahlende Pflanzen auf. Sehr stark behaarte Pflanzen werden oft als *C. lanatum* Lam. bezeichnet. Zwischen den beiden extremen Formen treten alle Übergänge auf und geographisch lassen sie sich nach Söllner (1954) nicht trennen. Nur eine zytotaxonomische Untersuchung der ganzen Gruppe kann abklären, ob und wie weit das vielgestaltige *C. alpinum* systematisch gegliedert werden kann."

Binz & Becherer (1976) do not consider *C. glomeratum* as a species with alpine distribution. They include C. *carinthiacum* Vest ssp. *austroalpinum* Kunz as a further species found in the southern part of the Alps.

Hartz (1895) distinguishes four subunits of *C. alpinum* L.: var. *legitima* (Lindbl.), var. *lanatum* (Lindbl.), var. *procerum* Lge., and var. *caespitosum* Malmgr.

Gelting (1934: 35) based on the identification of *C. alpinum* L. emended by Murbeck (1898) says that all transitional forms between f. *legitima* (Lindbl.) and f. *lanatum* (Lam.) Hegetschw. occur. *var. legitima* not found by him in East Greenland. The species *C. arcticum* is described by Lange (1880, cited from Gelting, 1934: 36): "(*C. arcticum*) was based on material from the west coast of Greenland between 64° and 70° N. lat. Lange claims the identy of his specimens with several forms from Scandinavia formerly described under the name *C. latifolium*, but at the same time he holds that they deviate from the alpine *C. latifolium*, and hence the species *C. arcticum* is erected. Murbeck (1898: 246) showed that *C. arcticum* is synonymous with *C. latifolium* var. *edmonstonii* Wats. Together with Ostenfeld he erected *C. edmonstonii* (Wats.) Murbeck et Ostf. The same year Ostenfeld determined Lange's type specimens of *C. arcticum* and some other specimens found in Herb. C. as *C. edmonstonii* (Wats.) Murbeck et Ostf., as appears from the labels in Herb. C. Later on, however, Ostenfeld (1926) must have relinquished this view, since in his paper: The origin of the flora of Greenland, 1926, neither *C. arcticum* Lge., nor *C. edmonstonii* (Wats.) Murbeck et Ostf. are mentioned. I have been unable to find anything in the literature about this altered view of Ostenfeld, but if we go through the material in Herb. C., we shall find that Ostenfeld has personally corrected all previous determinations of *C. arcticum* Lge. and *C. edmonstonii* (Wats.) Murbeck et Ostf. from Greenland to *C. alpinum* L. Thus Ostenfeld evidently held the view, that *C. edmonstonii* does not occur in Greenland, in which, from my present knowledge, I fully agree with him. Lange's specimens from Greenland beeing connected with the true *C. alpinum* by numerous transitional forms. The Scandinavian specimens denominated *C. arcticum* must thus be referred to *C. edmonstonii* or to *C. latifolium* var. *edmonstonii* Wats. if they are at all separable from *C. alpinum.*"

So, according to Gelting (1934) Ostenfeld has rejected *C. arcticum* Lge. as a Greenlandic species replacing this name on Lange's type specimens by *C. edmonstonii* (Wats.) Murbeck et Ostf. Later on he changed the name again on the labels to *C. alpinum* L.

It is interesting to see how other authors have treated this taxon. Seidenfaden (1933: 49) writes about *C. alpinum* L.: "Within this district, as throughout the whole of Greenland, the species is highly polymorphous. The relation between the various forms ist still far from elucidated, and deserve a close investigation. By application of the current terminology the majority of individuals should be referred to forma *lanatum* Lindb., forma *legitimum* Lindbl. being rare."

Seidenfaden & Sørensen (1937: 37) have chosen the name *C. alpinum* L. *s.l.* Holmen (1957: 43) uses the name *C. alpinum* L. and comments as follows: "Like many of the other species in Greenland this too belongs to a complex of species rich in variation, of which an acceptable account is still wanting. The variation of the species within Peary Land is very slight; all material belongs to a densely tufted and hairy form, with rather short and obtuse leaves."

Böcher (1959) distinguishes *C. alpinum* ssp. *lanatum* and *C. arcticum* in his study in Middle West Greenland. Accepting again the species *C. arcticum* Lge. he does not follow Ostenfeld (1926), who has dropped this name.

Schwarzenbach (1961) uses *C. alpinum* coll. based on the determinations of his collection from WA4 by K. Holmen.

Fredskild (1996 b) lists *C. alpinum* L. ssp. *lanatum* (Lam.) Asch. & Graebn. and *C. arcticum* Lge. for the region around Jakobshavn, using the same nomenclature as Böcher (1959). Fredskild (1966: 8/9) uses the name *C. arcticum* Lge. for his collection from Peary Land, commenting that Holmen (1957) had chosen *C. alpinum* L. for the same taxon. He says: "According to Hultén (1956) the North Greenland material can be referred to two closely related varieties, var. *vestitum* Hultén and var. *sordidum*. The former is the most common type in the area visited by the author. Hultén has determined two collections of var. *sordidum* loc. 11 John Murray, leg. Th. Wulff, and Navy Cliff Land at the head of Independence Fjord, leg. P. Freuchen."

Halliday et al. (1974: 11) refer their collections to *C. arcticum s.l.* Böcher et al. (1978: 147) present another taxonomic treatment of the *C. alpinum/arcticum*-complex based on morphological, cytological, ecological and geographical arguments. *C. alpinum* L. ssp. *lanatum* (Lam.) Asch. & Graebner is distinguished from *C. arcticum* Lge. with three subspecies: *C. arcticum* ssp. *arcticum* (syn. *C. edmonstonii* Murb. & Ostf.), ssp. *procerum* (Lge.) Böch., and ssp. *hyperboreum* (Tolm.) Böch. (syn. *C. hyperboreum* Tolm.). *C. alpinum* ssp. *lanatum* has 2n = 72 or 2n = 144 chromosomes. *C. arcticum* ssp. *arcticum* and ssp. *hyperboreum* have 2n = 108, ssp. *procerum* 2n = 54 oder 2n = 108.

C. alpinum ssp. *lanatum* differs morphologically from the others and has a different number of chromosomes. It prefers dry herb-slopes or "steppe" and is found south of ca. 78° in West Greenland and 74° in East Greenland (Bay 1992). The three subspecies of *C. arcticum s.l.* are differentiated by morphological characteristics, by their ecological demands, by their geographical distribution and by their chromosome numbers.

In the *C. alpinum/arcticum*-complex, Böcher et al. (1978) separate *C. alpinum* L. ssp. *lanatum* (Lam.) Asch. & Graebn. from *C. arcticum* Lge. *s.l.* Both species have high numbers of chromosomes and may be, therefore, poly-ploids of a primary type with 2n = 18. As is known from other examples, polyploid races often reproduce by self-polination, by apomictic modes of reproduction (e.g. species of *Rubus, Potentilla* or *Hieracium*) or by vegetative propagation. As a consequence a variety of local races may occur and can be considered taxonomically as subspecies, varieties or forms. Until the morphological and cytotaxonomical variation, the mode of reproduction and the ecological demands of these varieties are fully analysed it is best, to treat them as collective species (*coll.* or *s.l.*).

Scoggan (1978) treats *C. arcticum* on the level of varieties. Bay (1992: 14) writes: "The material of the *Cerastium alpinum/arcticum* complex has been treated as one taxon: *C. arcticum* Lge. as no convincing *C. alpinum* was found. The material from district North Greenland is very uniform whereas the collections in the southern part seem to be more related to *C. alpinum*."

Fredskild (1996 a) follows Böcher et al. (1978) suggesting that *C. alpinum* ssp. *lanatum* types are most frequent in the lowland areas towards the south of Greenland, whereas *C. arcticum* types prefer alpine localities. This agrees with the situation in South Greenland (Feilberg 1984, maps 65 and 62). He has analysed all the West Greenland material in Herb. C., summarizing the results of his study as follows: "The large W. Greenland material consists of c. 900 sheeets of which many match the description of the different taxa, and maps of these have been made. However, all too many indeterminable collections were left, including also some more well defined types e.g. a fairly glabrous type growing in willow scrubs, resembling *C. arcticum* but having white, 5-10 celled hairs on the basal leaves. Consequently, only one map, including all collections of the complex, is given."

In accordance with Bay (1992) and Fredskild (1996 b) the present material (83 specimens) is treated as one taxon *C. arcticum* Lge. *s.l.* while beeing aware that a few individuals could be considered as *C. alpinum* ssp. *lanatum.*

The analysis of this controversial discussion suggests the need for a detailed taxonomic study of the *C. arcticum/alpinum* complex in Greenland.

Cystopteris fragilis (L.) Bernh. *s.l.*

Böcher et al. (1978: 46) distinguish a southern ssp. *fragilis* (south of 68°35′ in East Greenland) and a northern ssp. *dickieana* (Sym.) Hyl. found north of

65°35′ in East Greenland. Following Bay (1992) and Fredskild (1996 b) the two subspecies are not separated, even if they can be sperated by spore sculpture ornamentation.

Draba adamsii Led. *s.l.*
Böcher et al. (1978: 114) consider *D. oblongata* R. Br. nom. ambig. as a synonym of *D.adamsii* Led., of *D. pauciflora* R. Br. nom. obs., and of *D. micropetala* Hook. According to Bay (1992: 14) *D. adamsii* and *D. oblongata* cannot be distinguished when flowers are absent. Consequently, he includes *D. oblongata* in *D. adamsii*. Following his view all specimens found in WA1, WA3-WA5 have been classified as *D. adamsii s.l.*

Draba alpina L. *s.l.*
Bay (1992: 15) discusses the *D. alpina*-complex with *D. alpina* L., *D. gredinii* Ekm. and *D. bellii* Holm. These three taxa are treated as separate species in Böcher et al. (1978: 114).

Bay (1992) includes *D. gredinii* Ekm. in *D. alpina,* saying that the colour of petals cannot be determined in dried material. This method of identification is followed with the exception of two specimens. These plants were in full flower and could be determined in the field without difficulties as *D. gredinii*.

Draba arctica J. Vahl *s.l.*
Schwarzenbach (1961) has referred *D. arctica* J. Vahl to *D. cinerea* Adams and (partially) to *D. arctogena/groenlandica* Ekm. *D. arctica* J. Vahl is a heterogenous species and is in need of a taxonomic revision.

Draba bellii Holm
Contrary to Scoggan (1978), and Böcher et al. (1978: 114), the two authors Bay (1992: 15) and Fredskild (1996 b) consider *D. alpina* and *D. bellii* as separate species. Fredskild bases his classification on specific differences in the shape and hairiness of the pods and the hairiness of the leaves. Further, he points out that the two taxa have different distribution areas: "(*D. bellii*) is a high arctic taxon. Towards its southern limit it mainly grows at higher levels. Thus, on Disko and Nuussuaq, 33 coll. are from 500 m, 9 from 200-500 and only 6 are from below this level, mainly found at rivers which might have brought seeds down. Contrary, *D. alpina* is a middle arctic taxon with lowland as well as alpine occurrences all over its distribution area."

The two species are separated in the present study.

Draba gredinii Ekm. (see *D. alpina* L. *s.l.*)

Draba lactea Adams / *D. fladnizensis* Wulf.
Fredskild (1996 b) discusses the separation of *D. lactea* and *D. fladnizensis* col-

lected in West Greenland. Based on his work collections intermediate between *D. lactea* and *D. fladnizensis* are excluded in the present study.

Dryas octopetala L. *s.l.* / *D. integrifolia* M. Vahl *s.str.*
Holmen (1957: 75-79) discusses the taxonomy of the Greenlandian species of *Dryas*, based on Juzepczuk (1929) and Porsild (1947). He identifies his material from Peary Land as *D. chamissonis* Spreng. ex Juz. and *D. integrifolia* M. Vahl. Holmen (1957) writes – in agreement with Porsild (1947) – that *D. punctata* Juz. is also present in Greenland. Hultén (1941-1950) treats *D. punctata* Juz. as a subspecies. Böcher et al. (1978: 69) distinguish the two species *D. integrifolia* M. Vahl and *D. octopetala* L. ssp. *punctata* (Juz.) Hult., but do not refer to *D. chamissonis* Spreng. ex Juz.

Elkington (1965) has studied the variation of the genus *Dryas* in Greenland. In his opinion the high variability of the Greenland *Dryas* is best explained by a process of introgression of the two species *D. integrifolia* and *D. octopetala* being the two extremes of a series of intermediate forms. In considering the circumpolar distribution of *Dryas*, *D. integrifolia* is a species of the western Arctic, while *D. octopetala* is an eastern species. According to Elkington (1965) it seems problablе that the introgression of the two species takes place over the whole range of distribution of *D. octopetala* in Greenland. Discussing the occurence of the N.E.Siberian species *D. crenulata* Juz. in Greenland, Elkington (1965) states that Porsild (1947) has probably confused the measurements of the leaves given by Juzepczuk (1929), giving leaf dimensions of 3.0-4.5 cm long and 0.6-1,4 cm broad instead of 0.6-1.4 cm long and 3-4.5 mm broad. He says that Holmen has based the description of *D. crenulata* on Porsild's publication. Elkington (1965: 18) considers a specimen found in WA4, Krumme Langsø and determined by Th. Sørensen as *D.* cf. *crenulata* Juz. as a large leaved modification of *D. octopetala*.

Bay (1992: 15) distinguishes the western *D. integrifolia* M. Vahl from the eastern *D. octopetala* L. He uses the following characters in distinguishing the two species: 1) the presence of ‚octopetala scales' (Elkington 1965) and 2) serration of the leaves. He states: "*D. octopetala* is defined by having typical "*octopetala* scales" and being serrated to the tip of the leaves. *D. integrifolia* is defined by lacking scales and lacking serrations or having only 1-2 at the basal part of the leaves. Other combinations of the characters are considered hybrids."

The present collection includes only a few specimens which are referred to as *D. integrifolia* M. Vahl *s.str.* Most plants are considered to be hybrids between the two species and are included in *D. octopetala* L. *s.l.*

Equisetum arvense L. *s.l.*
Based on the decription of the subspecies in Böcher et al. (1978: 40) and following Bay (1992: 23) this variable species is treated as *E. arvense* L. *s.l.*

Hippuris vulgaris L.
Bay (1992: 22) uses the name *H. vulgaris* L. *s.l.* as a collective species. As Böcher et al. (1978) do not mention any subspecies, the name *H. vulgaris* L. has been kept.

Melandrium affine J. Vahl *s.l.* / *Melandrium triflorum* (R. Br.) J. Vahl *s.str.*
Following Böcher et al. (1978: 160) and Bay (1992: 22) it seems best to treat the two taxa as *M. triflorum s.str.* and *M. affine s.l.*

Papaver radicatum Rottb. *s.l.*
Böcher et al. (1978: 101) state that three species are recognizable in the North Greenland part of the distribution of *P. radicatum* coll., namely *P. radicatum s.str.* and two undescribed specimens with 2n = 84 and with 2n = 70 respectively. Bay (1992: 16) writes: "Since it was not possible to recognize the two undescribed entities on the basis of herbarium material the taxon is treated here as a single species."

All specimens from WA1 – WA5 are included in *P. radicatum* Rottb. *s.l.*, as did also Bay (1992).

Phippsia algida (Sol.) R. Br. *s.l.*
Bay (1992, 1993) has revised the Greenland material of *P. algida* (Sol.) R. Br. and separates the two taxa *P. algida* ssp. *algida* and *P. algida* ssp. *algidiformis* (H. Sm.) Löve & Löve.

C. Bay and B. Fredskild have determined specimens from WA1, WA2, WA3 and WA5 as ssp. *algidiformis*. Dr G. Halliday (pers. comm.) has informed me that no specimen collected by the author in WA2 or WA4 and kept at the Lancaster Herbarium belong to *P. algida* ssp. *algidiformis*.

Poa alpina L. *s.l.*
As Böcher et al. (1978) state, that several chromosomal races of non viviparous *P. alpina* exist with chromosome numbers within the range of 2n = 14 to 2n = 74. *P. alpina* var. *vivipara* is an ecologically and geographically differing variety.

Schwarzenbach (1956) has shown by cultivation experiments that the mode of reproduction of a viviparous clone from Scoresby Sund depends on temperature conditions and photoperiodism. Reproduction started only after a period with temperatures below freezing. The plants produced flowering shoots under short-day conditions, and shoots with bulbils only when the day-light lasted longer than c. 14 hours. In the late spring and in autumn the plants developed shoots with intermediate types of inflorescences.

All specimens found in WA1 – WA4 are non viviparous. They are included in *P. alpina s.l.*

Potentilla hookeriana Lehm. *s.l.* / *Potentilla nivea* L. emend. Hultén *s.str.*
The complex *P. nivalis / P. hookeriana in Greenland* has been subject of taxonomic discussions for a long time. In consequence of the apomictic reproduction local races occur which are genetically uniform and often intermediate between the two taxa.

Gelting (1934: 104) comments *P. nivea* as follows: "As numerous varieties and forms have been erected under this polymorphous species, a subdivision must pay great regard to material collected in other Arctic regions, but this is beyond the scope of the present paper. I have attempted to classify my specimens, but the attempt has been unsuccessful, since it seems that practically all transitional forms occur". And: "The variability of *P. nivea* may partly be due to the habitat and partly to the history of the species. Its area of distribution must be supposed to have been divided into several small areas during the Quartenary deterioriation of the climate, viz.: one south of the Quartenary ice-shield, another on the Polar side of this shield, and further, possible unglaciated areas within the space of the ice-shield proper. The isolation may have brought about a differentiation of the originally more homogeneous species. After the close of the Ice Age the areas again unitized, but exhibit now a number of more or less distinguishable subspecies."

Seidenfaden & Sørensen (1937: 59) write: "*P. nivea* is of common occurrence and very rich in forms within the area. For the delimitation of the species, see Sørensen 1933: 61 f.; *Potentilla Pedersenii* RYDB. is referred to *P. nivea.*"

Holmen (1957: 80) follows Hultén (1941 – 1950) and includes all specimens of the *P. nivea* group from Peary Land under the name of *P. chamissonis* Hultén. He emphasis that there should be a new revision of the whole arctic *P. nivea* group based upon cultivation experiments and cytological investigation. He says: "No true *P. nivea* in the Hultén (1945) sense was found in Peary Land, while on the other hand the closely related *P. chamissonis* seems to be rather frequent here."

Fredskild (1966: 13) mentions only *P. chamissonis* from Peary Land. Böcher et al. (1978: 75) separate *P. hookeriana* Lehm. with ssp. *hookeriana* and ssp. *chamissonis* Hultén from *P. nivea* L. emend. Hultén. The northern species *P. hookeriana* is a very variable species. Several chromosome numbers are mentioned (2n = 56, 63, 28, 42, 49 and 77). The hybrid *P. hookeriana* x *P. nivea* has 2n = 63 chromosomes. The southern *P. nivea* is also a variable species. It reaches the northern limit in East Greenland at 76°21'N.

Bay (1992: 36) refers all *P. hookeriana* from North Greenland and North East Greenland to ssp. *chamissonis*. *Potentilla hookeriana s.l.* has in Greenland a northern distribution (map 50) while the southern *P. nivea s.str.* (map 95) appears to reach 78°N at the East coast.

Fredskild (1996 b) unites the ssp. *hookeriana* and ssp. *chamissonis* to *P. hookeriana s.l.* Accepting his arguments the collections from WA1 – WA5 are separated in *P. hookeriana s.l.* and *P. nivea s.str.*

Puccinellia bruggemanni Th. Sør.
P. bruggemannii is not mentioned in Böcher et al. (1978). Bay (1992, 1993) discusses this northern species and maps the distribution. Revising the collection from Kronprins Christian Land (WA5) he found one specimen of *P. bruggemannii* collected at an altitude of 650 m (05.08.1952).

Saxifraga hyperborea R.Br. (see *S. rivularis s.l.*)

Saxifraga nivalis L. / *Saxifraga tenuis* (Wbg.) H. Smith
Following Bay (1992), and Böcher et al. (1978) *Saxifraga nivalis* and *Saxifraga tenuis* are treated as two distinct species in this study.

Saxifraga rivularis L. *s.l.* / *Saxifraga hyperborea* R. Br.
Böcher et al. (1978: 89) have separated *S. rivularis* (2n = 52) from *S. hyperborea* (2n = 26). Bay (1992) has mapped *S. hyperborea* as a circumgreenlandic species (map 16), while *S. rivularis* does not reach North Greenland (map 98). Fredskild (1996 b), however, includes *S. hyperborea* in *S. rivularis s.l.* stating that there are intermediate forms between species. He follows Scoggan (1978/79) who treats *S. hyperborea* as a form. Fredskild (pers. comm.) has proposed that all specimens from WA1-WA5 should be included in the taxon *S. rivularis s.l.*

Saxifraga tenuis (Wbg.) H. Sm. (see *S. nivalis* L.)

Stellaria longipes Goldie *s.l.*
Böcher (1951 a) and Philipp (1972) have treated the Greenland material of the *Stellaria longipes* complex. Böcher et al. (1966) split the taxon into 6 species of lower taxonomic value: *S. longipes* Goldie *s.str.*, *S. edwardsii* R. Br. (syn. *S. ciliatosepala* auct.), *S. laxmannii* Fish., *S.crassipes* Hult., *S. monantha* Hult. and *S. laeta* Richards. Bay (1992: 17) discusses the distribution of these species and mapped the collective species *S. longipes* Goldie *s.l.*

The present collections (13 specimens) are all included in *S. longipes* Goldie *s.l.*

4. Altitude distribution of collected species

4.1 Database

The main database (WA1-WA5) is based on plant lists of 626 sites with descriptions of the geographical place, vegetation, and phenology of the plants. These lists have been transferred to a relational database: Each observation of a species at a given site is characterised by a set of 24 variables. The geographic units used by the author are defined in Tab. 6.
The database includes 9626 records called "observations", 3457 of them (36%) are validated by herbarium specimens. Most of these specimens are kept in the Greenland Herbarium of the University of Copenhagen (Herb. C.), a smaller number is deposited in the Arctic Herbarium of the University of Lancaster (U. K.).

The analysis in this study is based in general on the 3457 herbarium specimens collected by the author. Collections of other botanists from the same areas are excluded for the following reasons:

- Many botanists visiting the study areas WA1-WA5 have collected their specimens near the coast or at low altitudes.
- There is often no information on the altitude of the sites in the labels.
- Dr Geoffrey Halliday (Lancaster University) is preparing a comprehensive study on the geographical distribution of vascular plants (including dot maps) in East Greenland between 71°N and 76°N. He will take account of

Tab. 6. Definitions of geographic units used by the author

Geographic unit	Definition
Area	The word area is used as a general expression.
Study area	Area studied by the author. The five study areas are named as WA1-WA5.
Sub-area	Part of a study area (two sub-areas in WA1 and three sub-areas in WA5).
Province	Geographic units used by the statisticians in chapters 9 and 12. The provinces PA and PB correspond to the sub-areas of WA1. The study area WA2 is split into the three provinces PC, PD and PE. The provinces PF and PG cover WA3. Province PH is identical with study area WA4, and province PI with WA5.
Location	Small and topographically homogenous sub-unit.
Site	Place in the field where plants have been collected or observed.
Niche	Isolated site or a few isolated sites in the same location.

Tab. 7. Numbers of sites arranged within study areas and altitude bands

Altitude bands (m)	WA1	WA2	WA3	WA4	WA5	WA1-5
0 - 299	27	45	15	8	48	143
300 - 599	33	4	16	8	10	71
600 - 899	20	14	13	8	12	67
900 - 1199	7	21	21	6	1	56
1200 - 1499	2	11	10	6	–	29
1500 - 1799	–	6	1	–	–	7
Total	89	101	76	36	71	373

all available collections (including the author's specimens) from the whole area.

- Bay (1992) has already included all known collections for East and Northeast Greenland north of 74°N.
- By being based on personal observations and collections of the author this has the advantage that all records show the same comparable criteria.

4.2 Herbarium specimens (3457) database

Distribution of sites (Tab. 7), numbers of herbarium specimens (Tab. 8), and numbers of species (herbarium specimens) for altitude bands (Tab. 9) are shown within each of the five study areas and for the whole area (WA1-WA5).

Plants have been collected at 373 sites (Tab. 7). In WA4 only 36 sites have been visited. This low number is explained by the fact that the collections in the area, where the helicopter landed, have been combined into one site.

The altitude distribution of the sites within the study areas is rather irregular. This was because the excursions were planned by the geologist and followed places of geological rather than botanical interest. Therefore, the pattern of altitude distribution is not constant throughout the study areas:

- WA1 has the highest number of sites (33) in the band 300-599 m. This number is explained by the altitude of the base camp near the airstrip at Pingo Pass (410 m). The lowest place visited during the summer 1991 was in Schuchert Dal with an altitude of 120 m.
- The site distribution in WA2 has two peaks. These are 45 sites within the lowest band, and 21 sites within the band between 900 m and 1199 m. This distribution reflects the botanical activity around the base camps at the coast and during the many stops of a motorboat trip in the Forsblad Fjord.

Tab. 8. Numbers of herbarium specimens arranged in study areas and altitude bands

Altitude bands (m)	WA1	WA2	WA3	WA4	WA5	WA1-5
0 - 299	203	328	131	130	246	1038
300 - 599	302	26	248	116	43	735
600 - 899	291	87	215	155	97	845
900 - 1199	74	160	128	191	1	554
1200 - 1499	27	113	44	67	–	251
1500 - 1799	–	31	3	–	–	34
Total	897	745	769	659	387	3457

On the other hand many plants have been collected around the transit camps in the interior of the Stauning Alper and in the Højedal below Ismarken.

- The altitude distribution of the sites is fairly homogenous in WA3 showing a peak of 21 in the band 900-1199 m. This number is explained by the long time spent on the high plateau of Månesletten in South Andrée Land.
- The 36 sites of WA4 are evenly distributed. Here there are no sites below 200 m, as the base camps were installed at higher levels by a seaplane at Vibeke Sø (1952), and at Krumme Langsø (1956).
- The largest number of sites in WA5 was situated between 0-299 m, since the two base camps at Ingolf Fjord and Hekla Sund were situated at sea-level, and the camp at Centrum Sø at an altitude of 30 m.

The number of specimens decrease at higher altitudes. The vertical distribution of the specimens within the five study areas shows some lack of consistency as a result of the position of base camps and transit camps. Because of this, the band ‚300-599 m' is underrepresented in the study areas WA2, WA4 and WA5, and the band ‚600-899 m' in WA2. As a consequence, species which could be expected to reach their highest levels within these bands may not all have been collected, or have been found only a few times.

The last column of Tab. 9 shows that the number of collected species is

Tab. 9. Numbers of species arranged in study areas and altitude bands

Altitude bands (m)	WA1	WA2	WA3	WA4	WA5	WA1-5
0 - 299	89	95	67	77	86	160
300 - 599	96	25	97	71	35	135
600 - 899	84	62	84	75	47	130
900 - 1199	40	66	66	72	1	109
1200 - 1499	26	48	33	27	–	73
1500 - 1799	–	20	3	–	–	21
Number of species	126	120	128	123	93	176

gradually decreasing from low to high altitudes. Nevertheless, 109 of 176 species (62%) have been collected at altitudes between 900 m and 1199 m.

The differences between the study areas reflect the influence of the mean temperature during growing season. WA1 is exposed to the maritime influence of Scoresby Sund and Kong Oscar Fjord. WA2 and WA3 belong to the more continental zone of the middle and interior fjords exposed to the warm and dry föhn winds from the west. Although WA4 is situated more than 200 km further north than WA1, the pattern of vertical distribution of species is very similar. The continentality of WA4 compensates for the difference in latitude between the two study areas. In WA5 (80°N) vascular plants have only been collected in the first four bands because of the combined influence of the northern latitude and the close vicinity to the outer coast.

5. Patterns of altitude distribution

5.1 A system for the classification of altitude distribution types

Bay (1992) has introduced a system for the classification of the geographical distribution of species in northern Greenland (North Greenland distribution types [NGDT]). He has compared his types with the Greenland and holarctic distribution according to Hultén (1958) and with the biological distribution types proposed by Böcher (1959). These systems for the classification (see also Yurtsev 1994) help to understand the present patterns of geographical distribution. An attempt therefore has been made here to develop an additional system based on the altitude distribution of vascular plants.

5.2 Altitude distribution types (ADT): Methods of classification

This system of classification (ADT) of the observed 176 species has been developed by the author on an empirical base combining on three criteria:

- Preferred habitat
- Altitude distribution
- Latitude distribution between 71° – 81°N.

As a first step 103 common species have been classified according to the moisture of their habits into the six types A-F (Tab. 10).
The method of classification to altitude is based on the highest records of the 45 species of type B (Tab. 10) in 26 geomorphological regions corresponding to valleys, glaciers, nunataks or other conspicuous characteristics (Tab. 11). These species show a rather regular altitude distribution as the habitats

Tab. 10. Classification of habitats for 103 species

Type ADT	Number of species	Habitat
A	19	dry soils with a high content of mineral-salts, continental steppe
B	45	moderately moist soils, fell-field vegetation, heath
C	18	snow-patches, solifluction-soils, wet sand
D	14	mires
E	3	ponds and lakes
F	4	sea-shore, salt marshes

Tab. 11. List of regions in WA1-WA5 with the altitude of the highest sites

WA (n)	Name and position of the region
WA1 (2)	Mountains south of Pingo Pass (1090 m), Werner Bjerge north of Pingo Pass (1370 m)
WA2 (8)	*Stauning Alper:* Sefstrøm Gletscher (1670 m), Vikingebræ (870 m), Gully-gletscher (1690 m), Skjoldungebræ (1720 m), eastern side-glacier of Linné Gletscher (1335 m), western side-glacier of Linné Gletscher (1150 m) *Nathorst Land:* near Ismarken (1500 m), Schaffhauserdalen (1100 m)
WA3 (7) m),	*South Andrée Land:* Junctiondal (1310 m), plateau east of Junctiondal (1250 Månesletten (1285 m), Gauligletscher (1150 m) *East Andrée Land:* Lucia Dal, Benjamin Dal (1300 m) *North Andrée Land:* Moraenedal (1500 m), Agardh Bjerg (1500 m)
WA4 (5)	Waltershausen Gletscher/Waltershausen Nunatak (1270 m), North Strindberg Land/Studer Land (1200 m), Bartholin Land (1470 m), C. H. Ostenfeld Nunatak (760 m), Stenø Land/Vibeke Sø (900 m)
WA5 (4)	*Kronprins Christian Land:* Drømmebjerg (770 m), Centrum Sø (700 m) *Ingolf Fjord* (470 m) *Hekla Sund* (900 m)

Abbreviations
WA study areas WA1-WA5
(n) number of regions

"moderately moist soils, fell-field vegetation and heath" are evenly distributed between the seashore and the upper altitude limit. The distribution of the altitude limits in the 26 regions have been used to define empirically three altitude classes 1-3 for the five study areas (Tab. 12).

The altitude range for each area was divided into three classes and each species was scored according to its highest altitude, 1 being for the highest range and 3 the lowest. The five scores for each species were then averaged, to the nearest whole number, to produce an overall score and altitude class.

The ranges of the three classes differ between and within WA1-WA5. The classes 1 and 2 include mainly species with northern oder northeastern distribution in Greenland, while class 3 contains in general only the species reaching their northern limit south of 76°N, and therefore not found in WA5.

Using this scheme of classification an attempt was made to classify all species of the other types A, C – F. Based on the highest altitudes, each species is classified for each study area. The average of the obtained scores (1, 2 or 3) is used, as it is shown by two examples (Tab. 13).

Tab. 12. Classification of the three altitude classes in WA1-WA5 (altitude in meters)

	WA1	WA2	WA3	WA4	WA5
Class 1	1220-1000	1720-1450	1500-1100	1470-960	900-650
Class 2	999- 780	1449-1000	1099- 710	959-460	649- 0
Class 3	779- 0	1099-7100	709- 0	459- 0	-

Tab. 13. Two examples to show the method of classification (altitude classes)

Species		WA1	WA2	WA3	WA4	WA5
Calamagrostis purpurascens		690	1230	1285	960	330
	Class	3	2	1	1	3
Carex rupestris		770	1200	970	1180	670
	Class	3	2	2	1	1

The five scores for *Calamagrostis purpurascens* average 2.0, and for *Carex rupestris* it is 1.8. Therefore, both species are attributed to altitude class 2. The remaining 73 species have been defined as separate types, and are set out below:

- Type ADT G: 28 rare species with northern limits south of 76°N, found only in one two of the the four study areas WA1 – WA4.
- Type ADT H: 13 species found only in WA4 and WA5 (southern limit at 74°N)
- Type ADT X: 32 species with irregular patters of altitude distribution (not classified to altitude classes 1 – 3).

Based on this system of classification each species can by symbolised by a letter (A – H) followed by a numerical code 1, 2 or 3 for altitude class (except type X), indicating the altitude distribution. These categories are named "altitude distribution types ADT".

Tab. 14. Classification of all 176 collected species (ADT, and classes)

Type	n	%	sum	%
A1	8	4.5		
A2	11	6.3	19	10.8
B1	21	11.9		
B2	13	7.4		
B3	11	6.3	45	25.6
C1	5	2.9		
C2	7	4.0		
C3	6	3.4	18	10.3
D1	1	0.6		
D2	7	4.0		
D3	6	3.4	14	8.0
E3	3	1.7	3	1.7
F3	4	2.3	4	2.3
G3	28	15.8	28	15.8
H1	4	2.3		
H2	9	5.1	13	7.4
X	32	18.1	32	18.1
Sum	176	100.0	176	100.0

5.3 Patterns of altitude and geographical distribution of the altitude distribution types

Species of the main habitat types ADT A, B, C and D are found in at least three of the five study areas. Species of the type E ("water-plants") are recorded from WA2 and WA4. Plants of the type ADT F ("species of sea-shore") have been collected only in WA2, Forsblad Fjord. The distribution of ADT G is restricted to WA1-WA4. The type ADT H includes the species found in the northernmost study areas WA4 and WA5.

Species of the type ADT D ("mires") depend on special ecological conditions and are situated in low and middle altitudes. The 32 species of the type "X" could not be classified by the system with the types A – H. Their altitude and geographical distribution is discussed in the chapters 6 and 11.

6. WA1-WA5 Altitude distribution of species with data and comments

6.1 General remarks

A width of 300 m for the altitude interval (= band) has been chosen after several attempts with other intervals. Using this scale it is easy to visualize the pattern of altitude distribution of a species within a study area and to compare the situation between different study areas. The different bands are arranged in ascending order beginning at sea-level. The position of a band is given by the middle of the interval between the lower and the upper limit of the altitude band, e.g. "150 m" for the band "0-299" m. The median is the most accurate parameter for comparison between species within the same area and for comparison of an individual species in different areas. The altitude of the highest site where a species has been collected is noted in the column "max".

The species are arranged in the "Altitude Distribution Types (ADT)" in the following tables which have been described in chapter 5. Within these grouping (ADT) the species are in alphabetic order. Whenever an additional information for a species is given in another chapter, the reference follows the name of the species.

Tab. 15. Altitude distribution of all 176 species

Species	n	max	150	450	750	1050	1350	1650	ave	med
Alopecurus alpinus	21	1200	6	2	5	7	1	0	644	700
Antennaria canescens	18	1335	2	2	8	2	4	0	769	710
Antennaria porsildii	2	1010	0	0	1	1	0	0	*	*
Arabis alpina	6	770	1	1	4	0	0	0	545	620
Arabis holboellii	1	360	0	1	0	0	0	0	*	*
Arctagrostis latifolia	28	1000	10	11	5	2	0	0	413	497
Arctostaphylos alpina	17	650	9	7	1	0	0	0	280	260
Arenaria humifusa	1	240	1	0	0	0	0	0	*	*
Arenaria pseudofrigid.	12	1250	2	3	2	4	1	0	700	710
Armeria scabra ssp. *sibirica*	8	900	5	2	0	1	0	0	264	125
Arnica angustifolia	24	1230	8	4	8	3	1	0	535	575
Betula nana	21	870	12	7	2	0	0	0	275	200
Botrychium lunaria	1	770	0	0	1	0	0	0	*	*
Braya humilis	16	1160	9	5	1	1	0	0	353	235
Braya linearis	4	1170	1	2	0	1	0	0	538	390
Braya purpurascens	28	1285	8	8	4	6	2	0	589	510
Braya thorild-wulffii	8	650	6	1	1	0	0	0	205	90
Calamagrostis purpurascens	29	1285	10	6	5	5	3	0	564	480
Campanula gieseckiana s.l.	23	1335	4	6	5	5	3	0	660	650
Campanula uniflora	29	1720	2	4	7	8	6	2	921	900
Cardamine bellidifolia	16	1660	1	1	2	5	6	1	1047	1165

Species	n	max	150	450	750	1050	1350	1650	ave	med
Carex atrofusca	19	1310	11	3	3	1	1	0	377	240
Carex bicolor	1	75	1	0	0	0	0	0	*	*
Carex bigelowii.s.l.	29	1080	16	9	2	2	0	0	301	90
Carex capillaris s.l.	31	1030	12	10	4	5	0	0	446	400
Carex glacialis	4	1300	0	1	1	1	1	0	890	795
Carex glareosa	2	5	2	0	0	0	0	0	*	*
Carex lachenalii	6	620	3	1	2	0	0	0	375	365
Carex marina s.l.	2	550	1	1	0	0	0	0	*	*
Carex maritima	19	1200	8	2	6	2	1	0	476	480
Carex microglochin	2	230	2	0	0	0	0	0	*	*
Carex misandra	64	1670	14	20	13	15	1	1	584	515
Carex nardina	59	1720	9	11	13	14	8	4	801	850
Carex norvegica	5	1120	1	0	1	3	0	0	786	920
Carex parallela	5	100	5	0	0	0	0	0	47	25
Carex rariflora	5	230	5	0	0	0	0	0	137	160
Carex rupestris	53	1200	16	18	12	6	1	0	490	480
Carex saxatilis	27	800	11	11	5	0	0	0	361	380
Carex scirpoidea	36	1110	12	12	9	3	0	0	430	395
Carex stans s.str.	2	35	2	0	0	0	0	0	*	*
Carex supina ssp. *spaniocarpa*	19	1300	8	2	2	6	1	0	550	450
Carex ursina	1	1	1	0	0	0	0	0	*	*
Cassiope tetragona	29	1335	8	9	5	4	3	0	557	450
Cerastium arcticum s.l.	83	1680	14	11	30	13	13	2	748	700
Cerastium regelii ssp. *caespitosum*	9	700	5	1	3	0	0	0	290	70
Chamaenerion latifolium	48	1700	14	13	12	7	1	1	533	480
Cochlearia groenlandica	7	1000	4	2	0	1	0	0	337	250
Colpodium vahlianum	17	1210	5	3	6	2	1	0	537	600
Cystopteris fragilis s.l.	37	1470	9	7	7	7	7	0	691	720
Deschampsia brevifolia	4	70	4	0	0	0	0	0	55	60
Draba adamsii s.l.	11	1285	1	2	4	3	1	0	751	700
Draba alpina s.l.	22	1200	4	9	6	2	1	0	561	545
Draba arctica s.l.	61	1460	20	10	16	13	2	0	562	650
Draba arctogena	4	1370	0	0	0	3	1	0	1095	1165
Draba aurea	2	360	1	1	0	0	0	0	*	*
Draba bellii	32	1285	9	6	9	5	3	0	580	650
Draba crassifolia	2	620	0	0	2	0	0	0	*	*
Draba fladnizensis	15	1230	2	2	8	1	2	0	698	700
Draba glabella	63	1200	22	12	20	8	1	0	504	540
Draba gredinii	2	1200	0	0	0	1	1	0	*	*
Draba lactea	27	1450	4	6	10	4	3	0	689	670
Draba nivalis	24	1720	10	1	1	6	4	2	700	890
Draba subcapitata	29	1670	4	2	9	6	6	2	873	875
Dryas integrifolia s.str.	13	930	9	3	0	1	0	0	228	75
Dryas octopetala s.l.	41	960	16	10	11	4	0	0	436	415
Elymus hyperarcticus	3	35	3	0	0	0	0	0	33	35
Empetrum nigrum ssp. *hermaphroditum*	9	960	7	1	0	1	0	0	185	70
Epilobium arcticum	2	650	1	0	1	0	0	0	*	*
Equisetum arvense s.l.	21	760	10	3	8	0	0	0	371	440
Equisetum variegatum	17	1000	7	6	2	2	0	0	382	390
Erigeron compositus	2	1150	0	0	0	2	0	0	*	*
Erigeron eriocephalus	17	1450	1	2	4	8	2	0	887	885

Species	n	max	150	450	750	1050	1350	1650	ave	med
Erigeron humilis	24	1030	2	6	13	3	0	0	625	660
Eriophorum callitrix	15	700	11	3	1	0	0	0	222	200
Eriophorum scheuchzeri	23	1150	8	7	6	2	0	0	447	500
Eriophorum triste	37	1140	14	11	7	5	0	0	429	440
Euphrasia frigida	14	1230	7	2	1	2	2	0	493	310
Eutrema edwardsii	5	780	2	2	1	0	0	0	359	380
Festuca baffinensis	15	1285	2	3	3	4	3	0	771	720
Festuca brachyphylla	29	1460	8	4	6	9	2	0	636	750
Festuca hyperborea	15	1200	2	3	5	4	1	0	714	670
Festuca rubra s.l.	20	810	12	3	5	0	0	0	299	150
Festuca vivipara	21	1150	1	7	10	3	0	0	641	670
Gentiana detonsa	1	40	1	0	0	0	0	0	*	*
Gentiana tenella	5	770	0	2	3	0	0	0	560	650
Harrimanella hypnoides	2	1030	1	0	0	1	0	0	*	*
Hierochlo' alpina	22	1270	4	4	4	7	3	0	735	865
Hippuris vulgaris	2	380	1	1	0	0	0	0	*	*
Honckenya peploides var. *diffusa*	5	30	5	0	0	0	0	0	9	5
Huperzia selago ssp. *arctica*	13	1240	1	9	1	1	1	0	532	490
Juncus arcticus	8	520	5	3	0	0	0	0	227	150
Juncus biglumis	46	1200	14	10	14	6	2	0	540	530
Juncus castaneus	24	720	13	5	6	0	0	0	345	245
Juncus trifidus	1	80	1	0	0	0	0	0	*	*
Juncus triglumis	19	900	10	3	5	1	0	0	377	250
Kobresia myosuroides	33	1180	11	6	8	8	0	0	537	480
Kobresia simpliciuscula	19	900	10	4	4	1	0	0	355	250
Koenigia islandica	5	780	1	2	2	0	0	0	500	500
Lesquerella arctica	21	1170	8	6	4	3	0	0	452	480
Luzula arctica	28	1300	6	8	7	1	0	0	587	617
Luzula confusa	56	1500	5	15	13	12	10	1	771	735
Luzula spicata	21	1660	5	0	5	6	4	1	797	900
Lycopodium annotinum ssp. *alpestre*	2	270	2	0	0	0	0	0	*	*
Melandrium affine s.l.	60	1670	20	11	12	10	5	2	603	555
Melandrium apetalum ssp. *arcticum*	40	1400	9	7	11	12	1	0	640	965
Melandrium triflorum s.str.	14	945	8	1	2	3	0	0	356	135
Minuartia biflora	30	1030	9	8	11	2	0	0	464	510
Minuartia rossii	10	770	4	2	4	0	0	0	400	415
Minuartia rubella	47	1670	7	10	11	11	7	1	752	705
Minuartia stricta	5	510	2	2	0	0	0	0	293	320
Oxyria digyna	37	1150	9	9	16	3	0	0	509	620
Papaver radicatum s.l.	42	1700	3	4	9	13	10	3	950	980
Pedicularis flammea	30	1230	12	11	5	1	1	0	412	390
Pedicularis hirsuta	33	1150	7	9	13	4	0	0	566	640
Pedicularis lapponica	9	390	6	3	0	0	0	0	179	150
Phippsia algida s.l.	27	1285	6	6	7	5	3	0	627	620
Pinguicula vulgaris	3	100	3	0	0	0	0	0	63	80
Pleuropogon sabinei	2	200	2	0	0	0	0	0	*	*
Poa abbreviata	33	1500	3	3	7	11	8	1	914	1090
Poa alpina s.l.	37	1330	13	7	14	2	1	0	495	560
Poa arctica	51	1360	10	8	6	20	7	0	747	900
Poa glauca	86	1500	32	9	16	21	7	1	597	650

Species	n	max	150	450	750	1050	1350	1650	ave	med
Poa hartzii	6	1140	3	1	1	1	0	0	363	180
Poa pratensis s.l.	21	1285	14	4	2	0	1	0	261	100
Polygonum viviparum	58	1200	15	18	19	5	1	0	507	490
Potamogeton filiformis	1	230	1	0	0	0	0	0	*	*
Potentilla crantzii	11	810	2	3	6	0	0	0	529	650
Potentilla hookeriana s.l.	41	1335	11	5	10	10	5	0	659	785
Potentilla hyparctica	15	1350	5	0	5	3	2	0	645	780
Potentilla nivea	22	1460	5	2	4	7	4	0	750	690
Potentilla rubricaulis	14	1150	4	3	5	2	0	0	549	580
Puccinellia angustata	13	810	6	2	5	0	0	0	346	330
Puccinellia bruggemanni	1	600	0	0	1	0	0	0	*	*
Pyrola grandiflora	19	970	6	8	3	2	0	0	443	440
Ranunculus affinis s.l.	12	1200	4	2	0	4	2	0	653	720
Ranunculus confervoides	2	680	1	0	1	0	0	0	*	*
Ranunculus glacialis	7	1350	0	0	1	4	2	0	1037	960
Ranunculus hyperboreus	3	1150	1	0	0	2	0	0	750	900
Ranunculus nivalis	4	620	1	1	2	0	0	0	440	560
Ranunculus pygmaeus	15	1200	2	6	4	2	1	0	581	550
Ranunculus sulphureus	7	780	1	4	2	0	0	0	530	500
Rhodiola rosea	3	85	3	0	0	0	0	0	45	40
Rhododendron lapponicum	15	450	8	7	0	0	0	0	233	250
Rumex acetosella s.l.	11	450	8	3	0	0	0	0	118	20
Sagina caespitosa	1	480	0	1	0	0	0	0	*	*
Sagina intermedia	6	780	4	0	2	0	0	0	260	40
Salix arctica	80	1460	23	17	23	12	5	0	558	555
Salix herbacea	3	1030	1	1	0	1	0	0	490	420
Saxifraga aizoides	30	1000	12	10	7	1	0	0	388	390
Saxifraga caespitosa s.l.	36	1720	1	6	7	10	11	1	983	1060
Saxifraga cernua	60	1540	9	8	21	10	10	2	759	700
Saxifraga foliolosa	10	690	3	3	4	0	0	0	442	535
Saxifraga hieracifolia	4	520	2	2	0	0	0	0	303	310
Saxifraga nathorstii	15	900	8	3	3	1	0	0	357	240
Saxifraga nivalis	55	1540	14	9	11	10	10	1	692	670
Saxifraga oppositifolia	64	1670	12	10	22	11	7	2	715	700
Saxifraga platysepala	7	780	2	1	2	0	0	0	431	500
Saxifraga rivularis s.l.	6	1250	3	1	1	0	1	0	421	245
Saxifraga tenuis	1	1670	0	1	0	0	0	0	*	*
Sibbaldia procumbens	5	1030	0	0	2	3	0	0	926	1010
Silene acaulis	41	1700	7	10	14	5	4	1	653	680
Stellaria humifusa	2	5	2	0	0	0	0	0	*	*
Stellaria longipes s.l.	13	780	6	0	7	0	0	0	398	600
Taraxacum arcticum	10	1230	0	4	4	0	2	0	746	675
Taraxacum arctogenum	4	470	3	1	0	0	0	0	150	50
Taraxacum brachyceras	5	1030	0	1	3	1	0	0	724	770
Taraxacum phymatocarpum	26	1450	1	4	9	10	2	0	831	830
Taraxacum pumilum	2	200	2	0	0	0	0	0	*	*
Thalictrum alpinum	4	770	1	0	3	0	0	0	540	650
Tofieldia coccinea	8	400	5	3	0	0	0	0	170	100
Tofieldia pusilla	23	1030	13	7	1	2	0	0	305	250
Triglochin palustre	5	100	5	0	0	0	0	0	58	60
Trisetum spicatum	68	1720	11	10	27	11	7	2	721	690

Species	n	max	150	450	750	1050	1350	1650	ave	med
Vaccinium uliginosum ssp. *microphyllum*	32	1120	13	10	7	2	0	0	373	385
Viscaria alpina	3	810	2	0	1	0	0	0	348	150
Woodsia glabella	13	1450	3	3	3	3	1	0	616	690

Abbreviations

n	number of specimens	
max	altitude of the highest record	
150		number of specimens collected from 0- 299 m
450		number of specimens collected from 300- 599 m
750	median point of each band	number of specimens collected from 600- 899 m
1050		number of specimens collected from 900-1199 m
1350		number of specimens collected from 1200-1499 m
1650		number of specimens collected from 1500-1799 m
ave	average altitude	
med	median of altitudes	

6.2 Comments on the altitude distribution type ADT A, A1-A2

6.2.1 Altitude distribution type (and class) ADT A1

Comments on the individual species

Carex nardina ADT A1 (**11.3.2** N 8, N 9, N 13, N 16, N 17, N 18; **11.4.5**)

The circumgreenlandic *C. nardina* (Bay 1992) is common in all five study areas. It grows on dry open soils, in dry heaths and in the fell-field vegetation.

As a calciphyte the species tolerates a high content of mineral salts in the soil. Therefore, *C. nardina* has often been collected in the föhn-steppe of the

Tab. 16. Altitude Distribution Type (and class) ADT A1: Highest altitude collections in WA1-WA5

ADT	Species	WA1	WA2	WA3	WA4	WA5
A1	*Carex nardina*	1090	1710	1285	1270	650
A1	*Draba bellii*	1200	*	1285	1170	650
A1	*Draba nivalis*	1100	1720	1190	1170	35
A1	*Poa abbreviata*	1200	1500	1285	1470	650
A1	*Poa glauca*	1100	1500	1180	1350	650
A1	*Potentilla hookeriana s.l.*	1200	1335	1200	1180	650
A1	*Saxifraga oppositifolia*	1200	1670	1180	1470	700
A1	*Taraxacum phymatocarpum*	1200	1450	1180	1180	*

ADT	Altitude type
WA1-WA5	Study areas
*	No specimens available

Altitudes in m

continental belt of the interior fjords. The species is often associated with *Dryas octopetala s.l., D. integrifolia s.str., Carex rupestris* and *Kobresia myosuroides*.

The species is widely distributed from sea-level to 1710 m. Even in WA5 *C. nardina* has been found at an altitude of 650 m (Drømmebjerg, 80°15'N, 21°44'W).

It is worthwhile to present the list of the highest sites (above 1000 m) and to define the uppermost altitude zone reached by *C. nardina* (WA1, south of Pingo Pass: 1090 m, Biskop Alf Dal: 1040 m; WA2, Sefstrøm Gletscher: 1200 m, 1540 m, Gullygletscher: 1230 m, 1670 m, 1690 m, Skjoldungebræ: 1720 m, east of Linné Gletscher: 1335 m, west of Linné Gletscher: 1030 m, 1120 m; WA3, Månesletten: 1285 m, Gauligletscher: 1150 m, east of Junctiondal: 1220 m, Agardh Bjerg: 1120 m; WA4, Waltershausen Nunatak: 1270 m, north-western Bartholin Land: 1180 m).

Draba bellii ADT A1 (**11.3.2** N 10, N 15, N 17, N 18)
D. bellii is a northern species with an isolated occurrence in West Greenland (Bay 1992). The species has the southern limit in East Greenland around 70°N. It grows on dry to humid soils, often at places free of snow. It is a calciphilous species (Böcher et al. 1978).

D. bellii was not found in the Stauning Alper and in Nathorst Land, but it was very common in Kronprins Christian Land, reaching the altitude of 650 m at Drømmebjerg.

The species was collected in general below 750 m. However, there are collections from places which are several hundred meters above the next-highest sites (WA1, south of Pingo Pass: 1090 m, Randspids: 1200 m; WA3, Månesletten: 1285 m, east of Junctiondal: 1170 m, Morænedal 1100 m, 1280 m; WA4, Bernhard Studer Land: 1170 m).

Draba nivalis ADT A1 (**11.3.2** N 15; **11.4.6**)
D. nivalis has a southern distribution with the northern limit in East Greenland at c. 78°N. The species has often been collected in the inner continental parts (Bay 1992). *D. nivalis* grows frequently on dry places exposed to the wind (Böcher at al. 1978).

D. nivalis has been found in WA1-WA5. Two records are from Kronprins Christian Land: Centrum Sø (80°07'N, 22°16'W, 35 m) and Ingolf Fjord 80°33'N, 20°26'W, 10 m) thus shifting the northern limit of the species to 80°33'N. The species was collected in heaths, in alpine herb-slopes, and in early-melting snow-patches, sometimes also on moraines and on dry ground. It was found at all altitudes from the sea-shore to the altitude limit of vascular plants.

D. nivalis has reached altitudes above 1300 m in the crystalline areas of WA1 (Randspids: 1370 m) and of WA2 (Gully Gletscher: 1460 m, 1670 m,

Skjoldungebræ: 1300 m, 1320 m, 1720 m, Linné Gletscher: 1335 m). The highest records from WA3, South Andrée Land (1190 m) and WA4, Bernhard Studer Land (1170 m) are just below the limit of 1200 m .

Poa abbreviata ADT A1 (**11.3.2** N 17)
P. abbreviata is a northern species with an isolated area in West Greenland (Bay 1992). At the East coast the species is found as far south as Jameson Land (Böcher et al. 1978).

P. abbreviata is a typical species of the fell-field vegetation. It is well adapted to the severe conditions of the highest altitude and reaches the altitude limit of vascular plants. The species is found in the most remote sites and was collected on many nunataks in East and Northeast Greenland (see map in Bay 1992). Populations with different numbers of chromosomes have been found: 2n = 42 (28, 70), as it is stated by Böcher et al. (1978).

P. abbreviata was collected in all five study areas. The common species has reached high altitudes in WA1-WA3 (WA1, Randspids: 1200 m; WA2, Linné Gletscher: 1150 m, Højedal: 1500 m; WA3, Månesletten: 1285 m, Junctiondal: 1150 m, 1180 m, Gauligletscher: 1150 m). It was very common in the continental parts of the nunataks north of 74°N, where eight specimens were found at altitudes between 1170 m and 1470 m (Schwarzenbach 1961). In WA5 *P. abbreviata* reached 670 m at Drømmebjerg and 450 m north of Ingolf Fjord.

The general pattern of geographical and altitude distribution is quite unusual. *Poa abbreviata* was found in all mountainous parts of the study areas, even if these sites of high altitude were geographically separated by glaciers, valleys and fjords.

Poa glauca ADT A1 (**11.3.2** N 17)
P. glauca is a circumgreenlandic species missing only from the coastal area of North Greenland (Bay 1992). The species is found in the fell-field and on dry slopes (Böcher et al. 1978).

P. glauca is common in WA1-WA5 and has been collected up to high altitudes. The species has been found in the innermost continental belt from the Schuchert Dal in WA1 to the nunataks of WA4.

There are no specimens collected in the uppermost altitude zone near the limit of vascular plants. However, the species has been found in niches of high altitude (WA1, north of Pingo Pass: 1100; WA2, Ismarken: 1500 m, Sefstrøm Gletscher: 1230 m, east of Linné Gletscher: 1150 m, west of Linné Gletscher: 1150 m; WA3, east of Junctiondal: 1180 m, Gauligletscher: 1140 m, Benjamin Dal: 1150 m; WA4, Bernhard Studer Land: 1170 m, Waltershausen Nunatak: 1270 m, Bartholin Land: 1180 m, 1200 m, 1300 m, 1350 m). In WA5 the species has been collected up to 650 m (Drømmebjerg) and to 140 m north of Ingolf Fjord (80°34'N, 20°28'W).

Potentilla hookeriana s.l. ADT A1 (**11.3.2** N 16, N 17)
P. hookeriana s.l. is a northern species, missing in the coastal areas of North Greenland and preferring the continental parts of East Greenland (Bay 1992). It is a calciphyte and grows on dry loess and grus (Böcher et al. 1978).

The species is a common plant in WA1-WA5 and was found in middle and high altitude. *P. hookeriana s.l.* has reached a few high sites (WA1, Randspids: 1200 m; WA2, Linné Gletscher: 1335 m; WA3, Junctiondal: 1200 m; WA4, north-western Bartholin Land: 1180 m; WA5, Drømmebjerg: 650 m).

Saxifraga oppositifolia ADT A1 (**11.4.6**)
S. oppositifolia is a circumgreenlandic species (Bay 1992). It grows on open ground below snow-patches, in wet sites, in rock-fissures and prefers soils with limestone at high altitudes. The species flowers early. It is highly variable. Phenotypes with different growth-forms, of variable habit and with different shapes of flowers have been observed. It is possible that geographically separated ecotypes might be found in the mountains of East Greenland.

S. oppositifolia is one of the most common species in WA1-WA5. It is a typical species of the fell-field vegetation and reaches the altitude limit of vascular plants. The species was collected in most of the regions, where species may have survived the latest advances of the ice. (WA2, Sefstrøm Gletscher: 1540 m, Gully Gletscher: 1460 m, 1670 m; WA4, western Bartholin Land: 1350 m, 1470 m). The species was collected at 700 m in Kronprins Christian Land, Drømmebjerg.

Taraxacum phymatocarpum ADT A1 (**11.3.2** N 16)
T. phymatocarpum is a northern species (Bay 1992) which has its southern limit at Disko in West Greenland and at Scoresby Sund at the East coast. It grows on slightly humid soils, on stony terraces and in rocks (Böcher et al. 1978). *T. phymatocarpum* seems to be a calciphilous species and is also found on soils with a high content of mineral salts.

T. phymatocarpum was collected in WA1-WA4. The species was often found at middle and high altitude. There are several sites in WA1-WA3 at 1100 m or still higher (WA1, Randspids: 1200 m; WA2, Ismarken: 1450 m; WA3, South Andrée Land: 1150 m, 1160, 1160 m, North Andrée Land: 1100 m; WA4, north-western Bartholin Land: 1180 m). There are also six sites between 640 and 960 m in the nunataks of WA4.

6.2.2. Altitude distribution type (and class) ADT A2

General comments

There is a somewhat irregular pattern of the altitude distribution of the highest records (Tab. 17). As the species of ADT 2A prefer dry soils with a high

Tab. 17. Altitude distribution type (and class) ADT A2: Highest altitude collections in WA1-WA5

ADT	Species	WA1	WA2	WA3	WA4	WA5
A2	*Braya purpurascens*	1090	20	1285	1170	450
A2	*Braya humilis*	520	*	1160	200	*
A2	*Calamagrostis purpurascens*	690	1230	1285	960	330
A2	*Carex rupestris*	770	1200	970	1180	670
A2	*Carex supina* ssp. *spaniocarpa*	*	1120	1300	930	70
A2	*Draba fladnizensis*	850	1200	1230	1180	*
A2	*Dryas integrifolia s.str.*	480	*	240	930	450
A2	*Dryas octopetala s.l.*	950	960	700	930	*
A2	*Hierochloë alpina*	*	1230	1250	1270	140
A2	*Kobresia myosuroides*	690	1030	950	1180	200
A2	*Lesquerella arctica*	520	100	650	1170	65

ADT Altitude type
WA1-WA5 Study areas
* No specimens available
Altitude in m

content of mineral salts, the suitable sites are scattered, even in the continental belt of WA1 to WA4.

Comments on the individual species

Braya purpurascens ADT A2 (Chapter **11.3.2** N 10, N 15)

B. purpurascens is a northern species with a gap in its distribution along the northern part of the West coast (Bay 1992). The southern limit in the east is at Scoresby Sund. *B. purpurascens* is a calciphyte (Böcher et al. 1978).

The species is rather common in WA1 and WA3-WA5. *B. purpurascens* is typical of areas with limestone and dolomites, as well as for dry, clayey soils with a high content of mineral salts. Therefore, it has only occasionally been found in the Stauning Alper and in Nathorst Land.

B. purpurascens is often found in the continental föhn-steppe associated with *B. humilis, Lesquerella arctica, Calamagrostis purpurascens,* and *Kobresia myosuroides,* and is typical of the western continental belt between 73°N and 74°N. *B. purpurascens* often passes the level of 1000 m (WA1, south of Pingo Pass: 1090 m, north of Pingo Pass: 1000 m; WA3, Månesletten: 1285 m, Junctiondal: 1190 m, Morænedal: 1280 m; WA4, Bernhard Studer Land: 1170 m).

Braya humilis ADT A2

B. humilis occurs in East, Northeast and North Greenland (Bay 1992) north of 71°30′N (Böcher et al. 1978). It grows on dry soils.

The species has been found in WA1, WA3 and WA4. As the species prefers dry ground rich in mineral salts it is typical of the continental belt in the western part of Andrée Land and of Ole Rømer Land. In areas with calcareous and dolomitic sediments *B. humilis* might reach rather high altitudes (WA3, Junctiondal: 1160 m; WA4, C. H. Ostenfeld Nunatak: 640 m).

Calamagrostis purpurascens ADT A2 (**11.3.2** N 4, N 8, N 9, N 10, N 18; **11.4.5**)
This tricentric species (Bay 1992) is found as far south as Scoresby Sund. There are at least two races in Greenland with different numbers of chromosomes (2n = 56, 58) as it is stated in Böcher et al. (1978). The species grows on dry soils, in the steppe and in the fell-field vegetation.

C. purpurascens has been collected in all five study areas. It as a typical element of the continental föhn-steppe of the western continental belt. However, in contrast to *Braya purpurascens* and *Lesquerella arctica* the species has also been found in the Stauning Alper.

C. purpurascens is a species of low and middle altitude, but has been collected also at several sites above 1100 m (WA2, Sefstrøm Gletscher: 1200 m, 1230 m, Linné Gletscher: 1130 m; WA3, Månesletten: 1285 m, Junctiondal: 1140 m). There are three more sites in the nunataks of WA4 with altitudes of at least 900 m (WA4, nunatak in North Strindberg Land: 960 m, Bartholin Land: 930 m, Vibeke Nunatak: 900 m). The species has been collected in WA5 near Ingolf Fjord (80°33'N, 20°00'W, 330 m) and near Sæfaxi Elv (80°13'N, 20°48'W, 200 m).

Carex rupestris ADT A2 (**11.3.2** N 4, N 8, N 9, N 16, N 17, N 18)
C. rupestris is a tricentric species (Bay 1992), common in East and Northeast Greenland north of Angmagssalik (Böcher et al. 1978). It prefers dry heaths, steppe and rocks.

The species is quite common in all five study areas. It is a typical species of the vegetation-type, alpine meadows' together with *C. nardina* and *C. misandra*. As *C. rupestris* prefers dry soils (heath, föhn-steppe, rocks) exposed to wind it is often found in the föhn-valleys of the inner fjords.

The altitude distribution is similar to *Cassiope tetragona. C. rupestris* is a plant of low and middle altitude below 850 m with a few exceptions (WA2, Sefstrøm Gletscher: 960 m, 1200 m, Linné Gletscher: 1120 m; WA3, Junctiondal: 970 m; WA4, North Strindberg Land: 960 m, north-western Bartholin Land: 1180 m).

Carex supina ssp. *spaniocarpa* ADT A2 (**11.3.2** N 8, N 11; **11.4.5**; **11.4.8**)
C. supina ssp. *spaniocarpa* has been mapped by Bay (1992) as a southern species not reaching 80°N. As stated in Bay & Fredskild (1994) the northern limit it now 80°13'N, based on my specimen collected 1952 in WA5. The species is typical of the vegetation-type "alpine herb-slopes".

C. supina ssp. *spaniocarpa* has been found in WA2-WA5. As the species prefers dry sunny slopes on crystalline rocks it has often been found in the Stauning Alper (WA2) and in the western continental belt from WA3 and

WA4. In general *C. supina* ssp. *spaniocarpa* has been collected at low and middle altitude. A list of the highest sites shows an interesting pattern (WA2, western Linné Gletscher: 1120 m, Schaffhauserdalen: 1100 m, Ismarken: 990 m; WA3, Luciadal: 1300 m, Agardh Bjerg: 1140 m; WA4: Bartholin Nunatak: 900 m, Bartholin Land: 880 m, 930 m).

Draba fladnizensis ADT A2 (**11.3.2** N 4, N 11, N 16)
D. fladnizensis is a tricentric species (Bay 1992). It is found in East Greenland between 70°N and 78°N. The species grows often on rather dry soils (Böcher et al. 1978).

D. fladnizensis has been collected in WA1-WA4. It follows the continental belt from the southern Werner Bjerge to the nunataks of 74°N preferring middle altitudes. However, there are some sites at or above 750 m (WA1, south of Pingo Pass: 800 m, Schuchert Dal: 850 m; WA2, Sefstrøm Gletscher: 1200 m, Tærskeldal: 750 m; WA3, Luciadal: 1230 m; WA4, north-western Bartholin Land: 1180 m).

Dryas integrifolia s.str. ADT A2
D. integrifolia s.str. is a circumgreenlandic species common to northern and north-eastern Greenland (Bay 1992). South of 78°N *D. integrifolia s.str.* is rare preferring the inner continental belt. The species grows mostly in the fell-fields and in heaths on neutral to alkaline soils (Böcher et al. 1978).

The species was found in WA1, WA3 and WA4. It has been collected only occasionally in contrast to the hybrids *D. integrifolia* x *D. octopetala*. There are five records of *D. integrifolia s.str.* from the western continental belt only at altitudes below 500 m.

Dryas octopetala s.l. ADT A2 (**11.3.2** N 4, N 9; **11.4.5, 11.4.7**)
D. octopetala s.l. is a monocentric species and is found at the eastern coast of Greenland only. The distribution of the species lies between Scoresby Sund and c. 78°20′ N (Bay 1992). The species grows mostly in the fell-field vegetation and in heaths on neutral or alkaline soils (Böcher et al. 1978).

D. octopetala s.l. is a common plant in the study areas WA1-WA4. It is the dominant species of the dry heath on alkaline sediments. In the western continental belt the species is also found in zones of crystalline rocks where the soil is enriched with mineral salts.

D. octopetala s.l. is a species of low and middle altitude. Only four specimens have been collected above 900 m. (WA1, Biskop Alf Dal: 950 m, south of Pingo Pass: 910 m; WA2, Sefstrøm Gletscher: 960 m; WA4, Bartholin Land: 930 m). The vertical distribution below 900 m is fairly regular, because the species produces a huge number of seeds which are widely distributed by wind.

Hierochloë alpina ADT A2 (**11.3.2** N4, N 9, N 13, N 14; **11.4.8**)
The circumgreenlandic species *H. alpina* (Bay 1992) is common in East Greenland. It prefers dry and sunny slopes and is an important component of the *Cassiope*-heath. The species is also found in the föhn-steppe.

H. alpina has been collected in WA2-WA5 following the western continental belt from the Stauning Alper to Ole Rømer Land (74°N). The pattern of altitude distribution within this belt is shown by the list of sites at or above 900 m (WA2, Sefstrøm Gletscher: 1230 m, Ismarken: 945 m; WA3, east of Junctiondal: 1100 m, Agardh Bjerg: 1250 m; WA4, nunatak in North Strindberg Land: 960 m, Waltershausen Nunatak: 1270 m, Bartholin Nunatak: 900 m, Bartholin Land: 930 m, north-western Bartholin Land: 1180 m, Vibeke Nunatak: 900 m). There is also one single specimen from WA5, Ingolf Fjord: 80°34'N, 20°28'W, 140 m.

Kobresia myosuroides ADT A2 (**11.3.2** N 8, N 9, N 16; **11.4.5**)
K. myosuroides has a circumgreenlandic distribution (Bay 1992). This common species grows in dry heaths and in the steppe (Böcher et al. 1978) .

K. myosuroides has been collected in WA1-WA5. The species has an interesting distribution in the study areas WA1-WA4. As a typical species of the föhn-steppe it follows the western continental belt from Schuchert Dal in the south to the nunataks north of 74°N, where the species has been collected from low-levels to altitudes of 650 m (WA3, Moraenedal) and 1180 m (WA4, north-western Bartholin Land).

Conversely, *K. myosuroides* has been found at several high places in the Stauning Alper (WA2, Sefstrøm Gletscher: 960 m, north of Vikingebræ: 870 m, Linné Gletscher: 920 m, 1030 m). Finally, *K. myosuroides* has been collected in WA5 with the highest record at 200 m in the Sæfaxi Dal.

Lesquerella arctica ADT A2
L. arctica is a northern species (Bay 1992) with its southern limit in East Greenland at 70°17' N (Böcher et al. 1978). This species is a calciphyte and grows also on soils with a high content of mineral salt.

L. arctica was found in all five study areas. It is a typical species of the steppe and has often been collected in the föhn-valleys of the western continental belt between Schuchert Dal in the south and the nunataks north of 74° N (WA4). There is only one record of *L. arctica* from WA5 (Ingolf Fjord, 65 m).

This species was often found in the lowest altitude bands, as a list of the highest sites in WA1-WA3 shows (WA1, Pingo Pass: 520 m; WA3, Benjamin Dal: 510 m, Moraenedal: 650 m). In WA4 the species is rather common and has been collected above 600 m (Bernhard Studer Land: 1170 m, Bartholin Nunatak: 900 m, Bartholin Land: 880 m, 930 m, C. H. Ostenfeld Nunatak: 640 m).

6.3 Comments on the altitude distribution type ADT B, B1-B3

6.3.1 Altitude distribution type (and class) ADT B1

General comments

The maximum altitude of each species show remarkable differences between the five study areas. Many species of the group reach the altitude limit of vascular plants in one are or even in several areas (e.g. *Papaver radicatum s.l.*, *Saxifraga caespitosa s.l.* or *Saxifraga cernua*).

Comments on the individual species

Campanula uniflora ADT B1 (**11.3.2.** N 4, N 13; **11.4.8**)

C. uniflora has a nearly circumgreenlandic distribution missing only in the western part of North Greenland (Bay 1992). Towards the north it grows on dry herb-slopes. In the south the species is found at high altitude in the fell-field vegetation (Böcher et al. 1978).

Tab. 18. Altitude distribution type (and class) ADT B1: Highest altitude collections in WA1-WA5

ADT	Species	WA1	WA2	WA3	WA4	WA5
B1	*Campanula uniflora*	1200	1720	1400	1470	80
B1	*Carex misandra*	1200	1670	1140	1180	670
B1	*Cassiope tetragona*	1090	1335	1220	900	670
B1	*Cerastium arcticum s.l.*	1200	1680	1330	1470	670
B1	*Cystopteris fragilis s.l.*	1200	1460	1220	1470	330
B1	*Draba arctica s.l.*	1100	1460	1160	1270	650
B1	*Draba lactea*	510	1450	1140	960	670
B1	*Draba subcapitata*	1200	1670	1150	1350	700
B1	*Erigeron eriocephalus*	1100	1450	1180	1270	*
B1	*Festuca baffinensis*	1220	*	1285	1180	670
B1	*Festuca brachyphylla*	1100	1460	970	1170	30
B1	*Melandrium affine s.l.*	1090	1670	1220	1350	650
B1	*Minuartia rubella*	1200	1670	1285	1350	670
B1	*Papaver radicatum*	1200	1700	1500	1470	900
B1	*Poa arctica*	1200	1360	1140	1350	670
B1	*Potentilla nivea s.str.*	1200	*	1285	1170	*
B1	*Salix arctica*	1100	1460	1150	960	700
B1	*Saxifraga caespitosa s.l.*	1200	1720	1330	1470	670
B1	*Saxifraga cernua*	1200	1540	1500	1470	700
B1	*Saxifraga nivalis*	1200	1540	1350	1350	670
B1	*Trisetum spicatum*	1200	1720	1150	1170	670

ADT Altitude type
WA1-WA5 Study areas
* No specimens available
Altitudes in m

C. uniflora was rather common in WA1-WA4, but has been also found in WA5 at Danmark Fjord (80°34'N, 24°16'W, 80 m). The species has been collected generally at medium and high altitude.

Carex misandra ADT B1 (**11.3.2** N 16, N 17)
C. misandra is a northern species common in most parts of its area (Bay 1992). Böcher et al. (1978) state that the southern limit is at 66°N, where the species is found mainly at higher altitudes.

C. misandra has been collected in all five regions. *C. misandra* has been found at surprisingly high sites (WA1, Randspids: 1200 m, Biskop Alf Dal: 1000 m, 1040 m; WA2, Sefstrøm Gletscher: 960 m, Gullygletscher: 1670 m, Linné Gletscher: 1030 m; WA3, Junctiondal: 1140 m; WA4, North Strindberg Land: 960 m, Bartholin Land: 1180 m; WA5, Drømmebjerg: 670 m).

Cassiope tetragona ADT B1 (**11.3.2** N 10, N 13, N 17, N 18)
C. tetragona is a northern species missing in the southernmost parts of Greenland (Bay 1992). It is often found in the alpine part towards the southern limit (Böcher et al. 1978). *C. tetragona* is a dominant element of the heaths on dry to humid ground. It is snow-covered in winter.

C. tetragona was collected in all five study areas. In the western continental belt the species is not as common as in the middle part of the fjords. The species grows at low and middle altitude reaching the altitude limit in WA1-WA4 around 800 m. However, there are a few exceptions (WA1, south of Pingo Pass: 1090 m; WA2, Sefstrøm Gletscher: 960 m, Skjoldungebræ: 1260 m; WA3, east of Junctiondal: 1190 m, Agardh Bjerg: 1220 m; WA4, Bartholin Nunatak: 900 m). The highest site in WA5, Drømmebjerg, has an altitude of 670 m.

Cerastium arcticum s.l. ADT B1
C. arcticum s.l. has a circumgreenlandic distribution (Bay 1992). It belongs to the typical species of the fell-field vegetation. The species has a broad ecological range and is found as a minor component of many types of vegetation in middle and low altitude.

C. arcticum s.l. has been collected in WA1-WA5 from sea-level to the altitude limit of vascular plants. The species was most common in the altitude band "600-899 m" (Tab. 10). *C. arcticum s.l.* is abundant and wide-spread in the nunataks north of 74°N.

The highest sites of the species give a clear picture how the altitude limit of vascular plants changes from one region to the other in WA1-WA4 (WA1, south of Pingo Pass: 1090 m, south of Randspids: 1200 m; WA2, Sefstrøm Gletscher: 1260 m, Gullygletscher: 1460 m, 1670 m, Skjoldungebræ: 1260 m, Linné Gletscher: 1200 m; WA3, Månesletten: 1285 m, Gauligletscher: 1140 m, east of Junctiondal: 1160 m, Agardh Bjerg: 1330 m; WA4, Bernhard Studer

Land: 1170 m, Waltershausen Nunatak: 1270 m, Bartholin Land: 1180 m, 1300 m, 1350 m, 1470 m). *Cerastium arcticum s.l.* has been collected in WA5 up to 670 m (Drømmebjerg).

Cystopteris fragilis s.l. ADT B1

C. fragilis s.l. has a similar northern distribution to *Chamaenerion latifolium* and *Trisetum spicatum* (Bay 1992). It grows in fissures of rocks, in dry heaths, around bird-perches and near erratic boulders.

C. fragilis s.l. was found in WA1-WA5. There are several regions where the species has been collected at unusually high altitudes (WA1, north of Pingo Pass: 1000 m; WA2, Gullygletscher: 1460 m, Sefstrøm Gletscher: 1200 m, Linné Gletscher: 1150 m; WA3, Junctiondal: 1140 m; WA4, Bernhard Studer Land: 1170 m, Waltershausen Nunatak: 1270 m, western Bartholin Land: 1470 m).

Draba arctica s.l. ADT B1 (**11.3.2** N 15, N 17)

D. arctica s.l is a northern species with an isolated occurrence in West Greenland (Bay 1992). The southern limit is near Scoresby Sund. The species grows on dry soils with grus and on moraines (Böcher et al. 1978).

D. arctica s.l. has been collected in WA1-WA5. The species has a broad ecological range. It is often found on dry terraces and sunny slopes. *D. arctica s.l.* has also been observed in the föhn-steppe and occurs in screes and rocks. The species is rather common in the Stauning Alper and in the western continental belt.

D. arctica s.l. was found from sea level to the vegetation-line. There are a few sites in WA1-WA4 at high altitudes indicating possible niches reached by the species before the latest advance of the ice (WA1, north of Pingo Pass: 1100 m, south of Pingo Pass: 1090 m; WA2, Gullygletscher: 1460 m, Linné Gletscher: 1150 m, Schaffhauserdalen: 1100 m; WA3, east of Junctiondal: 1160 m; WA4, Bernhard Studer Land: 1170 m, Waltershausen Nunatak: 1270 m). *D. arctica s.l.* was common in Kronprins Christian Land. The highest site is Drømmebjerg (650 m).

Draba lactea ADT B1 (**11.4.9**)

D. lactea has nearly a circumgreenlandic distribution, missing only in Southeast Greenland (Bay 1992). In East Greenland the species is found north of 68°10'N (Böcher et al. 1978). The species grows in snow-patches and heaths as well as on terraces and moraines or in screes and rocks.

This species has been collected in WA1-WA5. *D. lactea* is a mountain plant reaching sometimes the altitude limit of vascular plants. It is best characterised as a species of fell-field.

There are a few remarkably high sites in WA2-WA4 belonging to the pattern of niches discussed in chapter 11.4.9. (WA2, Skjoldungebræ: 1320 m, east of Linné Gletscher: 1335 m, Ismarken: 1450 m; WA3, east of Junctiondal: 1100 m, 1140 m; WA4, North Strindberg Land: 960 m, Bartholin Nunatak: 900 m). The highest site in WA5 is Drømmebjerg (670 m).

Draba subcapitata ADT B1 (**11.3.2** N 15)
D. subcapitata is a bicentric northern species with an isolated occurrence in West-Greenland. It has a similar distribution to *D. bellii* (Bay 1992). *D. subcapitata* is found in East Greenland north of Jameson Land. It grows on dry soils, in grus and in rocks in sites free of snow in winter (Böcher et al. 1978).
This species has been collected in all five study areas. *D. subcapitata* is one of the most typical species of crystalline mountains in East Greenland preferring the highest altitude bands. The list of all specimens collected above 1000 m is long and surprising.

Erigeron eriocephalus ADT B1 (**11.3.2**, N 12, N 14, N 15)
E. eriocephalus has been found in four different areas of Greenland. The largest area is between Scoresby Sund and Peary Land (Bay 1992). This northern species grows in the rocks and on dry hills. It has been found mainly at high levels towards the south (Böcher et al. 1978).

This species has been collected in WA1-WA4. It has been observed also in Kronprins Christian Land, but there are no specimens available. *E. eriocephalus* follows the western continental belt. The species belongs to the vegetation-type ‚fell-field'. It has not reached the altitude limit of vascular plants.

Festuca baffinensis ADT B1 (**11.3.2**, N 16)
F. baffinensis is a bicentric northern species (Bay 1992). Böcher et al. (1978) set the southern limit of *F. baffinensis* to the region of Scoresby Sund. The species grows in river-beds, in heaths, in snow-patches and in fell-fields.

F. baffinensis has been collected in the sedimentary parts of WA1, WA3-WA5, and has not been found in WA2. The species prefers middle and high altitude. *Festuca baffinensis* has been collected in several regions at high levels where immigrants from the north have probably survived the latest advance of the ice (WA1, south of Randspids: 1200 m, 1220 m; WA3, Månesletten: 1285 m, Gauligletscher: 1150 m, east of Junctiondal: 1140 m; WA4, north-western Bartholin Land: 1180 m).

Festuca brachyphylla ADT B1 (**11.3.2**, N 17)
F. brachphylla is a circumgreenlandic species missing only from the coastal part of North Greenland (Bay 1992). The species has a broad ecological range. It is found in snow-patches, in herb-slopes as well as on dry terraces and in fell-fields.

F. brachyphylla has been found in all five study areas. It was rather common in WA2 and WA4. The species prefers middle and high altitude without reaching the altitude limit of vascular plants. However, there are several specimens from altitudes at or above 900 m. (WA1, north of Pingo Pass: 1100 m; WA2, Sefstrøm Gletscher: 1200 m, Gullygletscher: 1460 m, west of Linné Gletscher: 1020 m, 1030 m, Højedal: 900 m; WA3, Moræenedal: 970 m; WA4, North Strindberg Land: 960 m, Bernhard Studer Land: 1170 m, Bartholin Land: 900 m, 930 m).

Melandrium affine s.l. ADT B1
M. affine s.l. is a northern species missing in the coastal areas of North Greenland (Bay 1992). It is found in East Greenland between Scoresby Sund and Danmark Fjord. The species grows on dry to moderately humid soils rich in nutrients (Böcher et al. 1978).

M. affine s.l. was quite common in WA1-WA5, often growing on dry soils and found on screes and rocks. As the species tolerates a rather high content of mineral salts in the soil it has been found in the föhn-steppe of the western continental belt from 72°N to 74°N.

M. affine s.l. reaches high altitudes but has not been collected at the altitude limit of vascular plants, even the species grows at rather high altitudes in WA2-WA4. (WA2, Sefstrøm Gletscher: 1540 m, 1670 m, Linné Gletscher: 1150 m, Ismarken: 990 m; WA3, east of Junctiondal: 1170 m, 1220 m, Agardh Bjerg: 1220 m; WA4, western Bartholin Land: 1200 m, 1350 m, Waltershausen Nunatak: 1270 m).

Minuartia rubella ADT B1
"Dry to wet soil sand and grus". This is the description of the sites where the circumgreenlandic species *M. rubella* might occur (Böcher et al. 1978).

M. rubella is common in WA1-WA5. This species has been found at high altitude, but was not collected at the altitude limit of vascular plants. There are numerous sites at high altitudes in several regions (WA1, north of Pingo Pass: 1200 m; WA2, Sefstrøm Gletscher: 1260 m, Gullygletscher: 1670 m, east of Skjoldungebræ: 1280 m, west of Linné Gletscher: 1150 m, Schaffhauserdalen: 1100 m; WA3, Månesletten: 1285 m, east of Junctiondal: 1150 m, Moræenedal: 1240 m; WA4, Bartholin Land: 1180 m, 1350 m).

Papaver radicatum s.l. ADT B1 (**11.3.2** N 10)
P. radicatum s.l. is a well-known and very common species with a circum-green-landic distribution (Bay 1992). Towards the south the species has often been recorded from high altitudes (Böcher et al. 1978).

P. radicatum s.l. was collected in all five study areas. Near the altitude limit it is often found in the vegetation type ‚fell-field‘. At lower levels the species is found in many vegetation types. The species was rather common in the highest altitude bands. Just 50% of the 42 specimens have been collected at altitudes of 1000 m or more. Poppies with white or yellow flowers have been observed in all study areas and at all altitudes.

The highest sites of *P. radicatum s.l.* are found in granitic or gneissic mountains. There is a more or less closed area of high sites in valleys with glaciers of the Stauning Alper (WA2, Sefstrøm Gletscher: 1260 m, 1540 m, Gullygletscher: 1460 m, Skjoldungebræ: 1700 m). There are three regions in Andrée Land (WA3, east of Junctiondal: 1150 m, 1310 m, Gauligletscher: 1150 m, Morænedal: 1500 m). In the nunatak region north of 74°N the species has been collected eight times at altitudes above 1150 m. The highest site is in western Bartholin Land: 1470 m. There are also three high sites from WA5 (Drømmebjerg: 700 m, west of Hekla Sund: 900 m, north of Ingolf Fjord: 330 m).

Poa arctica ADT B1 (**11.8.2** N 18)
P. arctica is a common grass with a circumgreenlandic distribution (Bay 1992). It grows in heaths and in the fell-field (Böcher et al. 1978).

This species is common in all five study areas. It was found at high altitude but was not collected at the altitude limit of vascular plants. The highest sites of *P. arctica* in WA1-WA4 belong to niches where immigrants from a time with a warmer climate may have survived the latest advance of the ice (WA1, south of Randspids: 1200 m; WA2, Sefstrøm Gletscher: 1030 m, Skjoldungebræ: 1360 m, east of Linné Gletscher: 1335 m, west of Linné Gletscher: 1120 m, Schaffhauserdalen: 1100 m, Ismarken: 945 m; WA3, Gauligletscher: 1140 m, Agardh Bjerg: 1100 m; WA4, Waltershausen Nunatak: 1270 m, Bartholin Land: 1180 m, 1350 m).

Potentilla nivea s.str. ADT B1
P. nivea s.str. is a species with a southern distribution (Bay 1992). It is found in Northeast and East Greenland with the northernmost site at Langelv, Hochstetter Forland, 75°12'N, 19°45'W. (Fredskild & Bay 1990). The species grows in dry herb-slopes and on rocks.

P. nivea s.str. has been collected in WA1-WA4. There are a few of unusual high sites (WA1, south of Randspids: 1200 m; WA2, Sefstrøm Gletscher: 1010 m, Gullygletscher: 1460 m, west of Linné Gletscher: 1120 m, Ismarken: 1450 m; WA3, Månesletten: 1285 m; WA4, Bernhard Studer Land: 1170 m, Bartholin Land: 900 m, 930 m).

Salix arctica ADT B1 (**11.4.7**)
S. arctica is a northern species (Bay 1992). It is very common and, it is found in many types of vegetation.

S. arctica is the most common species in the five study areas. It was collected at rather high altitudes but did not reach the altitude limit of vascular plants. The sites are in general evenly distributed from the lowest to the highest levels. However, there are a few high sites in WA1-WA3 which are separated from the next highest sites by gaps of more than 150 m. These sites belong to the regions where species might have survived the latest advance of ice (WA1, north of Pingo Pass: 1100 m; WA2, Sefstrøm Gletscher: 1200 m, 1260 m, Gullygletscher: 1460 m, Skjoldungebræ: 1320 m, east of Linné Gletscher: 1335 m; WA3: east of Junctiondal: 1150 m.

In WA4 there are eleven sites with the highest site in North Strindberg Land: 960 m. In Kronprins Christian Land *S. arctica* has been collected at many sites in all four sub-areas.

S. arctica produces high numbers of seeds carried by wind to high altitudes and over great distances. This high capacity for dispersal and ability to regenerate vegetatively explains the even distribution in all altitudes below 1000 m.

Saxifraga caespitosa s.l. ADT B1 (**11.3.2** N 10)

S. caespitosa s.l. is a circumgreenlandic species (Bay 1992). As a species of the fell-field vegetation it grows in fissures and on ledges of rocks, on dry slopes and in river-beds.

S. caespitosa s.l. was collected in WA1-WA5. Preferring granites and gneisses the species was particularly common in the Stauning Alper. *S. caespitosa s.l.* is a typical species of high altitude. It belongs to the plants which are best adapted to the severe conditions near the altitude limit of vascular plants. If the sites above 1100 m are mapped, the resulting pattern includes most of the niches where species may have survived the latest advance of the ice (WA1, south of Randspids: 1200 m; WA2, Sefstrøm Gletscher: 1260 m, Gullygletscher: 1460 m, east of Skjoldungebræ: 1720 m, west of Linné Gletscher: 1150 m, east of Linné Gletscher: 1335 m; WA3, east of Junctiondal: 1250 m, Gauligletscher: 1150 m, Agardh Bjerg: 1330 m; WA4, Bernhard Studer Land: 1170 m, Waltershausen Nunatak: 1270 m, Bartholin Land: 1180 m, 1300 m, 1330 m, 1470 m).

Saxifraga cernua ADT B1 (**11.4.7**)

The circumgreenlandic *S. cernua* (Bay 1992) is a common species found at sites with enough water during the time of early growth. It often develops the shoots rather late in snow-patches or below long-lasting snow-beds. Ptarmigan often disperse the bulbils of the species (Gelting 1937).

S. cernua was common in WA1-WA5 preferring the sedimentary parts. The species is typical for ‚fell-fields'. It is found at all altitudes and reaches the altitude limit of vascular plants.

S. cernua is evenly distributed in WA1, southern Werner Bjerge with the highest site from south of Randspids (1200 m). The species has been collected at high altitudes in the Stauning Alper (WA2, Sefstrøm Gletscher: 1540 m, Gullygletscher: 1230 m). There are three high sites from Andrée Land (WA3, Månesletten: 1285 m, Morænedal: 1440 m, 1500 m). Ten specimens have been collected in WA4; three sites are above 1200 m (Waltershausen Nunatak: 1270 m, western Bartholin Land: 1300 m, 1470 m). In WA5 *S. cernua* has frequently been collected in the sub-areas Centrum Sø, Ingolf Fjord and Hekla Sund. The highest site is Drømmebjerg: 700 m.

Saxifraga nivalis ADT B1 (**11.4.9**)
S. nivalis is a common circumgreenlandic species (Böcher et al. 1978, Bay 1992). It grows in snow-patches as well as on slopes and in rock fissures with enough water at the beginning of the growth season. The species prefers the crystalline parts of the study areas.

S. nivalis was found in all five study areas and was more or less evenly distributed between sea level and high altitudes. The species has not been collected in the uppermost band below the altitude limit of vascular plants. The species was found in most niches of high altitude where immigrants may have survived the latest advance of the ice (see chapter 11.4.9). *S. nivalis* is a good indicator of these niches (WA1, south of Randspids: 1200 m; WA2, Sefstrøm Gletscher: 1260 m, 1540 m, Gullygletscher: 1230 m, Skjoldungebræ: 1260 m, west of Linné Gletscher: 1100 m, east of Linné Gletscher: 1300 m; WA3, east of Junctiondal: 1150 m, Gauligletscher: 1150 m, Agardh Bjerg: 1350 m; WA4, North Strindberg Land: 960 m, Waltershausen Nunatak: 1270 m, Bartholin Land: 1180 m, 1350 m; WA5, Drømmebjerg: 700 m).

Trisetum spicatum ADT B1 (**11.3.2** N 15, N 17; **11.4.7**)
T. spicatum has a circumgreenlandic distribution with the exception of the coastal areas in North Greenland (Bay 1992). It grows in snow-patches, along wet slopes and also in the fell-field vegetation (Böcher et al. 1978).

The species was found in all five study areas. The highest sites indicate the niches where species might have survived the latest advance of the ice (WA1, north of Pingo Pass: 1100 m, south of Randspids: 1200 m; WA2, Sefstrøm Gletscher: 1200 m, 1540 m, Gullygletscher: 1230 m, Skjoldungebræ: 1310 m, 1320 m, 1720 m, east of Linné Gletscher 1335 m, west of Linné Gletscher: 1100 m, Ismarken: 1450 m; WA3, east of Junctiondal: 1150 m, Gauligletscher: 1150 m). There are also a few high sites in the nunataks north of 74°N (WA4, Bernhard Studer Land: 1170 m, nunatak in North Strindberg Land: 960 m, Bartholin Land: 880 m, 900 m; 930 m). The species was found at five sites in WA5 with a maximum of 670 m (Drømmebjerg).

6.3.2 Altitude distribution type (and class) ADT B2

General comments

The type (and class) B2 includes 13 species. Only 8 species reach Kronprins Christian Land north of 80°. There are a few species found at high altitudes These extreme values are discussed in the subchapter "Comments on the individual species".

Comments on the individual species

Arnica angustifolia ADT B2 (**11.3.2** N 3, N 4, N 13; **11.4.8**)
A. angustifolia has a tricentric distribution in Greenland (Bay 1992). The

Tab. 19. Altitude distribution type (and class) ADT B2: Highest altitude collections in WA1-WA5

ADT	Species	WA1	WA2	WA3	WA4	WA5
B2	*Arnica angustifolia*	840	1230	1050	900	*
B2	*Carex capillaris s.l.*	510	1030	720	930	*
B2	*Chamaenerion latifolium*	850	1700	850	1170	3
B2	*Draba glabella*	770	1200	1140	1170	*
B2	*Melandrium triflorum s.str.*	*	945	*	930	410
B2	*Pedicularis flammea*	70	1230	640	900	200
B2	*Pedicularis hirsuta*	1040	870	1150	900	75
B2	*Poa hartzii*	*	*	1140	640	330
B2	*Polygonum viviparum*	1200	960	720	960	75
B2	*Potentilla rubricaulis*	680	910	1150	*	470
B2	*Silene acaulis*	840	1700	720	930	75
B2	*Vaccinium uliginosum* ssp. *microphyllum*	770	1120	650	780	*
B2	*Woodsia glabella*	770	1450	720	930	*

ADT Altitude type
WA1-WA5 Study areas
* No specimens available
Altitudes in m

species is very variable. There are several races with different numbers of chromosomes (2n = 57, 76, 95) as is mentioned in Böcher et al. (1978). The species is found in dry sites, on herb-slopes, in screes, or around bird-perches.

A. angustifolia is widely distributed in the study areas WA1-WA4. The species prefers crystalline rocks, and it is therefore rather common in the Stauning Alper. *A. angustifolia* is a typical species of the mountain herb-slopes and grows below early melting snow-patches with favourable exposure even at rather high altitudes (WA2, Sefstrøm Gletscher: 960 m, 1230 m; WA3, Moræneda1: 1050 m; WA4, Bartholin Nunatak: 900 m).

Carex capillaris s.l. ADT B2 (**11.3.2** N 4, N 8; **11.4.8**)
C. capillaris s.l. is found south of 80°N with the exception of one record from North Greenland (Bay 1992). The species grows on wet rocks, heaths, in fens and on clay (Böcher et al. 1978).

C. capillaris s.l. has been collected in WA1-WA4, but was missing in Kronprins Christian Land. *C. capillaris s.l.* is well represented in the continental belt running from the Schuchert Dal northwards to Alpefjord and further north to Andrée Land and to the nunataks of Ole Rømer Land.

This species has a similar altitude distribution to *Dryas octopetala s.l.*, with only three sites at or above 900 m (WA2, Linné Gletscher: 1030 m; WA4, Bartholin Land: 930 m, Bartholin Nunatak: 900 m).

Chamaenerion latifolium ADT B2 (**11.3.2** N 6, N 7, N 8, N 15; **11.4.7**)
Bay (1992) classifies *C. latifolium* together with *Cystopteris fragilis* and *Trisetum*

spicatum as ‚NGDT 1c (circumgreenlandic, but missing in some coastal areas of North and Northeast Greenland)'. The same three species reach their vertical limits below the typical species of ‚fell-field' vegetation. *C. latifolium* is found in river-beds, on flats with gravels as well as on screes and rocks. The seeds might easily be carried by wind or snow drifts over great distances and to high altitudes. It colonises disturbed sites.

C. latifolium was collected in all five study areas. The species is common in low and middle altitude, decreasing in number at higher levels. Some specimens have been collected at or above 900 m (WA2, Sefstrøm Gletscher: 960 m, Gullygletscher: 1230 m, Skjoldungebræ: 1700 m, west of Linné Gletscher: 1100 m; WA4, nunatak in North Strindberg Land: 960 m, Bernhard Studer Land: 1170 m, Bartholin Nunatak: 900 m, Bartholin Land: 930 m, Vibeke Nunatak: 900 m).

Draba glabella ADT B2 (**11.3.2** N 4, N 5, N 8, N 9, N 15;)
D. glabella is a bicentric species found in East Greenland between 69°N and 77°N (Bay 1992). It grows on dry slopes, in herb-slopes, in copses and heaths (Böcher et al. 1978).

D. glabella has been collected in WA1-WA4 in low and middle altitude with a few exceptions at sites above 800 m (WA2, Sefstrøm Gletscher: 960 m, 1010 m, 1200 m, north of Vikingebræ: 810 m, Linné Gletscher: 1150 m, Schaffhauserdalen: 1100 m, Højedal: 850 m; WA3, east of Junctiondal: 1140 m; WA4, Bernhard Studer Land: 1170 m, Bartholin Land: 880 m, 900 m, Vibeke Nunatak: 900 m).

Melandrium triflorum s.str. ADT B2
M. triflorum s.str. is a northern species (Bay 1992). Its southern limit is at the East coast in Scoresby Sund. There is also an isolated occurrence near Mt. Forel, 66°46'N (Böcher et al. 1978). *M. triflorum s.str.* grows on dry soils, on dry clayey terraces and in the fell-field.

This species has been collected in four regions of WA2, WA4 and WA5 restricted to middle altitude. As it is difficult to distinguish the two taxa *M. affine* and *M. triflorum* (see chapter 3) further studies are required to verify the altitude and geographical distribution of the two species.

Pedicularis flammea ADT B2 (**11.3.2** N 4, N 5; **11.4.8**)
P. flammea has been mapped as southern species found south of 80°N (Bay 1992). Böcher et al. (1978) set the northern limit at 78°N in Northeast Greenland. Bay & Fredskild (1994) mention now a new record from an area north of 80°: Campanuladal 80°38'N. The species grows in wet heaths or near running water. Sites of *P. flammea* are covered with snow during the winter.

This species was found in all five study areas WA1-WA5. In WA5 the only specimen was collected above Sæfaxi Elv in Kronprins Christian Land

(80°13′N, 20°48′W, 200 m). *P. flammea* has been found in middle altitude. Three specimens have been collected above 800 m (WA2, Sefstrøm Gletscher: 1230 m, north of Vikingebræ: 870 m; WA4, Bartholin Land: 900 m). These three sites belong to niches at rather high altitudes.

Pedicularis hirsuta ADT B2 (**11.3.2** N 2, N 10)
P. hirsuta is a northern species found at the East coast as far south as Angmagssalik (Böcher et al. 1978). The species is most common in dwarf-shrub heaths and is in general snow-covered during the winter. It has more or less the same altitude distribution as *Vaccinium uliginosum* ssp. *microphyllum*.

P. hirsuta is a common species in the sedimentary parts of all five study areas. It requires soils with enough humidity during the growth-season. Therefore, the number of specimens collected in the föhn-valleys of the western continental belt is small. In WA5 the species was found only around Centrum Sø at altitudes from 40 to 75 m.

Most specimens have been found at altitudes below 900 m with some exceptions (WA1, Biskop Alf Dal: 1040 m; WA3, Gauligletscher: 1150 m; WA4, Bartholin Land: 900 m).

Poa hartzii ADT B2
P. hartzii is distributed in East and North Greenland (Bay 1992). The species is found in the beds of rivers, on clayey soil and on dry slopes (Böcher et al. 1978).

This species was found in WA3-WA5 at low and middle altitude with only one exception (WA3, east of Junctiondal: 1140 m)

Polygonum viviparum ADT B2 (**11.3.2** N 1; **11.4.7**)
The species *P. viviparum* is a common species all over Greenland and is found in many types of vegetation as snow-patches, heaths and herb-slopes (Böcher et al. 1978). As the bulbils are often carried away by birds, especially by ptarmigan (Gelting 1937), the species may reach sites at rather high altitudes. An example is a snow-patch in WA1, south of Randspids (1200 m) where feeding birds have been observed (03/08/1991).

P. viviparum is a common plant in the study areas WA1-WA4. In WA5 it has been collected only twice around Centrum Sø . *P. viviparum* grows in general at low and middle altitude. The highest sites in WA2-WA4 are below 1000 m (WA2, Sefstrøm Gletscher: 960 m, north of Vikingebræ: 870 m; WA3, Morænedal: 720 m; WA4, North Strindberg Land: 960 m, Bartholin Land: 880 m, 900 m).

Potentilla rubricaulis ADT B2 (**11.3.2** N 9, N 10, N 18; **11.4.6**)
P. rubricaulis is found in North and East Greenland and has a small area in West Greenland (Bay 1992). Scoresby Sund is the southern limit in East

Greenland (Böcher et al. 1978). The species is found on solifluction-soils, on grus and around bird-perches.

This species was collected in WA1-WA3 and in WA5. The southernmost site was in WA1, south of Pingo Pass (71°49'N, 24° ,W, 690 m). This site is a link to the next occurrence in Jameson Land, Ubaliq, south of Gurreholm Bjerge (Fredskild et al. 1982). Two specimens are from Nathorst Land (WA2, Schaffhauserdalen: 160 m, Højedal: 910 m). A well established population has been found in South Andrée Land with the highest site near Gauligletscher: 1150 m. There are two sites in WA5 (Sæfaxi Dal: 50 m, north of Ingolf Fjord: 470 m).

Silene acaulis ADT B2 (**11.3.2** N 4, N 6, N 7, N 8)
S. acaulis is a circumgreenlandic species missing only in the coastal areas of North Greenland (Bay 1992). It is a species of the fell-field (Böcher et al. 1978).

This species was found in all five study areas, but was rare in WA5. It is common in WA1, reaching an altitude of 840 m east of Pingo Pass. There are several sites of high altitude in the Stauning Alper. (WA2, Sefstrøm Gletscher: 1030 m, 1200 m, 1260 m, Gullygletscher: 1230 m, 1460 m, north of Vikingebræ: 810 m, Skjoldungebræ: 1700 m, Linné Gletscher: 1100 m). The species has been collected in WA3 at low and medium altitude (Benjamin Dal: 650 m, Morænedal: 720 m). *Silene acaulis* has been found at several sites on the nunataks north of 74°N (WA4, Waltershausengletscher: 450 m, Bartholin Land: 640 m, 900 m, 930 m, C. H. Ostenfeld Nunatak: 640 m). Two collections are from WA5, Centrum Sø: 40 m, 75 m.

Vaccinium uliginosum ssp. *microphyllum* ADT B2 (**11.3.2** N 4, N 8; **11.4.8**)
The species *V. uliginosum* ssp. *microphyllum* does not extend north of 82°N (Bay 1992). In East Greenland it is a common and typical plant of the heath.

V. uliginosum ssp. *microphyllum* has been collected in WA1-WA4. It is a species of middle altitude, occasionally found above 700 m (WA1, Biskop Alf Dal: 770 m, Sefstrøm Gletscher: 960 m, west of Linné Gletscher: 1120 m; WA3, Moraenedal: 720 m; WA4, Vibeke Sø: 780 m). It has also been collected in the nunataks north of 74°N (WA4, Waltershausen Gletscher: 450 m).

Woodsia glabella ADT B2 (**11.3.2** N 8, N 9)
W. glabella is a circumgreenlandic species, missing only in the coastal areas of North Greenland (Bay 1992). It grows often in fissures of rocks and prefers calcareous soils (Böcher et al. 1978).

W. glabella was found in WA1-WA4. Most sites are below 700 m with the exception of a few niches where species might have survived the latest advance of the ice (WA1, Biskop Alf Dal: 770 m, WA2: Sefstrøm Gletscher: 960 m, west of Linné Gletscher: 1030 m, Ismarken: 1450 m, WA3: Moraenedal: 720 m, WA4, Bartholin Land: 930 m).

6.3.3 Altitude distribution type (and class) ADT B3

General comments

The type (and class) ADT B3 contains 12 species, found in WA1-WA4, but missing in WA5. The four species *Carex scirpoidea, Festuca rubra s.l., Gentiana tenella* and *Vaccinium uliginosum* ssp. *microphyllum* have been collected in a sheltered niche at the unusual altitude of 770 m (WA1, Biskop Alf Dal). Specimens of *Carex scirpoidea, Euphrasia frigida, Tofieldia pusilla* and *Vaccinium uliginosum* ssp. *microphyllum* have been found in altitudes above 1000 m in WA2.

Comments on the individual species

Arctostaphylos alpina ADT B3

The southern species *A. alpina* has a tricentric distribution (Bay 1992) and is found at the East coast between 68°45'N and 74°30'N (Böcher et al. 1978). The species grows in dry heaths and on dry sites exposed to the sun.

A. alpina has been found in WA1-WA4. The species prefers dry and sunny heaths and has often been observed in the more continental parts of the four study areas. The species has only been collected in the three lowest altitude bands (WA1, Pingo Pass: 510 m; WA2, north of Vikingebræ: 220 m; WA3, Morænedal: 650 m; WA4, Vibeke Sø: 450 m).

Betula nana ADT B3 (**11.3.2** N 5; **11.4.8**)

B. nana is a southern species with the limit at c. 76°N in East Greenland (Fredskild and Bay 1990). The species grows in dry to wet heaths and also in the fell-field vegetation of low altitude.

Tab. 20. Altitude distribution type (and class) ADT B3: Highest altitude collections in WA1-WA5

ADT	Species	WA1	WA2	WA3	WA4	WA5
B3	*Arctostaphylos alpina*	510	220	650	450	*
B3	*Betula nana*	540	870	700	350	*
B3	*Campanula gieseckiana*	350	1335	650	900	*
B3	*Carex scirpoidea*	770	1110	650	350	*
B3	*Euphrasia frigida*	520	1230	400	200	*
B3	*Festuca rubra s.l.*	770	810	*	240	*
B3	*Pedicularis lapponica*	390	*	85	350	*
B3	*Rhododendron lapponicum*	340	380	440	450	*
B3	*Rumex acetosella s.l.*	*	300	400	450	*
B3	*Tofieldia coccinea*	320	380	400	*	*
B3	*Tofieldia pusilla*	490	1030	100	60	*

ADT Altitude type
WA1-WA5 Study areas
* No specimens available
Altitudes in m

B. nana was collected in four study areas WA1-WA4. It is fairly common in the dry valleys of the continental belt of the middle and interior fjords, and is often dominant in dry heaths. The species grows at low altitude (highest sites: WA1, Pingo Pass: 540 m, Pingo Dal: 390 m; WA2, north of Vikingebræ: 870 m, Sefstrøm Gletscher: 380 m; WA3, Agardh Bjerg: 700 m, Benjamin Dal: 440 m; WA4, Vibeke Sø: 350 m).

Campanula gieseckiana s.l. ADT B3 (**11.3.2** N 4, N 7, N 8; **11.4.8**)
C. gieseckiana s.l. has a similar distribution in Greenland to *Betula nana* (Bay 1992). It grows in rocks and in fell-fields, in dry grassy meadows and on herb-slopes.

This species has been collected in WA1-WA4. It is quite common in the Stauning Alper, and has been found on steep sunny slopes or in sheltered fissures of rocks. The highest sites in WA2 are above 1000 m (WA2, Sefstrøm Gletscher: 1010 m, 1200 m, Skjoldungebræ: 1320 m, Linné Gletscher: 1120 m, 1335 m). The altitude limits of the species are much lower in WA1, Pingo Dal: 350 m; WA3, Morænedal: 650 m; WA4, Bartholin Land: 880 m, 900 m).

Carex scirpoidea ADT B3 (**11.3.2** N 4, N 5; **11.4.8**)
C. scirpoidea has a southern distribution. The northern limit in East Greenland is at 74°22'N (Böcher et al. 1978). It grows in well developed heaths, in copses with willow and on herb-slopes.

C. scirpoidea was collected in WA1-WA4. It has been found in dry to moist heaths and has a similar altitude distribution as *Betula nana*. *C. scirpoidea* occurs mostly below 800 m. However, there are four exceptions from the Stauning Alper (WA2, Sefstrøm Gletscher: 960 m, Vikingebræ: 870 m, Skjoldungebræ: 960 m, 1110 m).

Euphrasia frigida ADT B3 (**11.3.2** N 4, N 5; **11.4.8**)
E. frigida is a southern species with the northern limit in East Greenland at c. 77°N (Bay 1992). It grows on herb-slopes, in grassy meadows and also on ledges of rocks (Böcher et al. 1978).

E. frigida has been found in WA1-WA4. The species has a similar altitude distribution as *Betula nana* and is in general restricted to altitudes below 550 m. However, it has been found several times at higher levels in the Stauning Alper (WA2, Sefstrøm Gletscher: 1010 m, 1200 m, 1230 m, north of Vikingebræ: 810 m). These four sites belong to a region which might have been a refugium for southern species during the latest advance of the ice.

Festuca rubra s.l. ADT B3 (**11.3.2** N 5)
The southern species *F. rubra s.l.* has been found in East Greenland as far north as Hochstetter Forland (Böcher et al. 1978). It grows in open herb-slopes, and on sand and grus.

This species has been collected in WA1, WA2 and WA4 in the three lowest altitude bands only. It was fairly common in the sub-area WA1 around Pingo Pass.

Pedicularis lapponica ADT B3
P. lapponica is a bicentric species (Bay 1992) found in East Greenland between 69°N – 74°30'N (Böcher et al. 1978).

P. lapponica was collected in WA1, WA2 and WA4. The species has a similar altitude distribution to *Betula nana.* It was found in the western continental belt between Schuchert Dal (WA1) in the south and Vibeke Sø (WA4) in the north. The highest sites are below 400 m (WA1, south of Pingo Dal: 390 m, Schuchert Dal: 340 m; WA4: Vibeke Sø: 350 m).

Rhododendron lapponicum ADT B3
R. lapponicum is a southern species (Bay 1992). It reaches its northern limit in East Greenland at 77°39'N (Böcher et al. 1978). The species is typical of dry heaths, calcareous sediments and grows on soils with a high content of mineral salts.

R. lapponicum has been collected in WA1-WA4 in the two lowest altitude bands with the highest sites below 450 m (WA1, Schuchert Dal: 340 m; WA2, Sefstrøm Gletscher: 380 m; WA3, Benjamin Dal: 440 m; WA4, Vibeke Sø: 450 m).

Rumex acetosella s.l. ADT B3
R. acetosella s.l. is a southern species (Bay 1992). The northern limit at the East coast is 75°16'N (Böcher et al. 1978, Fredskild & Bay (1990). The species likes dry and sunny slopes, and it is found also in grassy ‚meadows' of low altitude.

R. acetosella s.l. has been collected in the continental parts of WA2-WA4. It was found several times along the coast of Forsblad Fjord. The species has been observed a few times at medium altitude up to 450 m (WA2, Tærskeldal: 300 m; WA3, Morænedal: 400 m; WA4: Waltershausengletscher: 450 m).

Tofieldia coccinea ADT B3
The tricentric species *T. coccinea* (Bay 1992) is found in East Greenland between Kangerlussuaq (68°16'N) and Skærfjord (77°32'N). The species grows in *Cassiope*-heaths, sometimes also in fell-field vegetation of middle altitude.

T. coccinea was collected in WA1-WA3 at altitudes from sea-level to 400 m (WA1, Schuchert Dal: 380 m; WA2, Sefstrøm Gletscher: 380 m; WA3: Morænedal: 400 m).

Tofieldia pusilla ADT B3 (**11.3.2** N 4; **11.4.8**)
T. pusilla is a species with a southern distribution (Bay 1992). The northern limit is at 74°30'N (Böcher et al. 1978). The species grows in wet heaths and mosses.

T. pusilla has been collected in WA1-WA4 at altitudes below 400 m with a few exceptions (WA1, Pingo Pass: 490 m; WA2, Sefstrøm Gletscher: 960 m, 1030 m, Sedgwick Gletscher: 780 m; WA3, Benjamin Dal: 440 m, Moraenedal: 400 m). There are also two sites from WA4 (Krumme Langsø: 200 m, Vibeke Sø: 350 m).

6.4 Comments on the altitude distribution type ADT C, C1-C3

6.4.1 Altitude distribution type (and class) ADT C1

General comments

All six species have been collected in WA1-WA4 and at least once at altitudes above 1000 m. The highest sites in WA5 are between 670 m and 700 m. The species does not reach the altitude limit of vascular plants.

Snow-patches and snow-beds often show an irregular pattern of distribution within a given area. They are often found on plateaus with active polygon-fields and in the lee of ridges and crests. The type (and class) ADT C1 is sparsely represented in the crystalline parts of WA2.

Comments on the individual species

Cardamine bellidifolia ADT C1
C. bellidifolia is a circumgreenlandic species (Bay 1992) often found at high levels in the southern parts of Greenland (Böcher et al. 1978). The species grows in snow-patches and in the fell-field.

Tab. 21. Altitude distribution type (and class) ADT C1: Highest altitude collections in WA1-WA5

ADT	Species	WA1	WA2	WA3	WA4	WA5
C1	*Cardamine bellidifolia*	*	1660	1400	1350	670
C1	*Luzula arctica*	1000	990	1150	1300	670
C1	*Luzula confusa*	1200	1335	1500	1350	670
C1	*Melandrium apetalum* ssp. *arcticum*	1100	*	1150	1400	670
C1	*Saxifraga tenuis*	620	1230	1150	1180	700

ADT Altitude type
WA1-WA5 Study areas
* No specimens available
Altitudes in m

C. bellidifolia has been collected in WA2-WA5. It has also been observed once in WA1, Pingo Pass: 500 m, but the specimen was lost later. The species was found at middle and high altitudes. There are 10 sites in WA2-WA4 above 1100 m (WA2, Skjoldungebrae: 1260 m, 1660 m; WA3: east of Junctiondal: 1150 m, Gauligletscher: 1150 m, Agardh Bjerg: 1400 m; WA4: Waltershausen Nunatak: 1270 m, Bartholin Land: 1180 m, 1200 m, 1300 m, 1350 m).

Luzula arctica ADT C1 (**11.4.7**)
L. arctica is a northern species (Bay 1992). The southernmost record from the East coast is at 66°30'N (Böcher et al. 1978). *L. arctica* grows in the *Cassiope*-heath and in wet places with mosses as well as in snow-patches and in wet polygon-fields.

This species has been collected in all five study areas. In general, *L. arctica* is a plant of low and middle altitudes (Tab. 10). But, there are some interesting exceptions. In WA1, north of Pingo Pass: 1000 m, the species was found in a constantly irrigated polygon-field. As this place is visited by geese it is quite possible that this locality is a pioneer-site (11.4.7). *L. arctica* was collected at several sites at or above 900 m in WA2-WA4 (WA2, Højedal: 990 m; WA3, east of Junctiondal: 1150 m; WA4, nunatak in North Strindberg Land: 960 m, Bartholin Land: 900 m, 1300 m).

Luzula confusa ADT C1 (**11.3.2** N 10)
L. confusa is a circumgreenlandic species (Bay 1992). The species has a broad ecological range and it is found as a minor component of many types of vegetation. However, near the vegetation-line it is a typical species of the high-altitude snow-patches and "constantly irrigated fields".

L. confusa was found in all five study areas WA1-WA5. It is common in middle and high altitudes. Several specimens were found at high altitudes well above the next highest sites and in geographically isolated regions (WA1, south of Randspids: 1200 m; WA2, Sefstrøm Gletscher: 1260 m, Gullygletscher: 1230 m, Skjoldungebræ: 1260 m, Linné Gletscher: 1335 m; WA3, east of Junctiondal: 1310 m, Gauligletscher: 1140 m, Moranedal: 1500 m; WA4, Waltershausen Nunatak: 1270 m, Bartholin Land: 1350 m).

L. confusa has immigrated to altitudes above 1100 m in the valleys with glaciers of the southern Werner Bjerge, of the Stauning Alper and of East Andrée Land. The species is also present at high levels in the nunataks north of 74°N.

Melandrium apetalum ssp. *arcticum* ADT C1
M. apetalum ssp. *arcticum* is a bicentric northern species (Bay 1992). It has been found southwards to c. 69°N and has been recorded towards the southern limit often at higher altitudes. The species grows in fissures of rocks, in fell-field vegetation and on herb-slopes (Böcher et al. 1978).

M. apetalum ssp. *arcticum* has been found in the sedimentary parts of WA1,

WA3-WA5. It is a plant of low and middle altitudes, but it was collected also above 1000 m growing in snow-patches and on ,constantly irrigated slopes' (WA1, north of Pingo Pass: 1100 m; WA2, east of Junctiondal: 1150 m, Gauligletscher: 1150 m; WA4: Bartholin Land: 1180 m, 1400 m). Two collections from WA5 are of interest (Drømmebjerg: 670 m, Hekla Sund: 15 m).

Saxifraga tenuis ADT C1
S. tenuis is a circumgreenlandic species (Bay 1992). The species is found at wet places, often in snow-patches and on solifluction-soils (Böcher et al. 1978). *S. tenuis* was observed in WA1 at several sites around Pingo Pass: 500-620 m. It has been collected in WA2-WA5 only occasionally (WA2, Gullygletscher; 1230 m; WA3, east of Junctiondal: 1150 m; WA4, nunatak in North Strindberg Land: 960 m, Vibeke Sø: 780 m, Bartholin Land: 900 m, 1180 m; WA5, Drømmebjerg: 670 m).

6.4.2 Altitude distribution type (and class) ADT C2

General comments

The type (and class) ADT C2 is only partially represented in WA2 and WA5. Calciphytes as *Colpodium vahlianum* or *Eutrema edwardsii* have not been found in WA2. Most sites are below 1000 m. The exceptions (WA1-WA3) are discussed in the comments on the individual species.

Comments on the individual species

Carex maritima ADT C2 (**11.3.2** N 1, N 10; **11.4.7**)
C. maritima is widely distributed in Greenland missing only in the coastal area of North Greenland and in the northern part of the West coast (Bay 1992). *C. maritima* grows in places with sand or grus, in river-beds and on so-called

Tab. 22. Altitude distribution type (and class) ADT C2: Highest altitude collections in WA1-WA5

ADT	Species	WA1	WA2	WA3	WA4	WA5
C2	*Carex maritima*	1200	*	970	900	40
C2	*Colpodium vahlianum*	1000	*	1210	780	600
C2	*Minuartia biflora*	840	1030	700	960	*
C2	*Oxyria digyna*	1000	100	1150	640	700
C2	*Ranunculus affinis s.l.*	*	1200	1200	200	65
C2	*Ranunculus hyperboreus*	*	*	1150	900	200
C2	*Ranunculus sulphureus*	680	*	*	780	250

ADT Altitude type
WA1-WA5 Study areas
* No specimens available
Altitudes in m

‚strandbredder'. It is also found amongst mosses if there are enough mineral salts (Böcher et al. 1978). Sites offering these ecological conditions are rare in the crystalline mountains of the Stauning Alper.

This species has been collected in WA1, WA3-WA5. It was rather common in WA3 and WA4. In WA5 *C. maritima* has been found at two sites around Centrum Sø: 40 m, 30 m.

C. maritima was collected mainly at altitudes below 700 m with a few exceptions (WA1, south of Randspids: 1200 m; WA3, east of Junctiondal: 970 m, Moræneda: 700 m; WA4, Bartholin Land: 900 m, C. H. Ostenfeld Nunatak: 700 m, 720 m). The presence of *C. maritima* in the continental belt between 73°N and 74°N is explained by the occurrence of soils with a high content of mineral salts in the föhn-exposed valleys.

Colpodium vahlinanum ADT C2 (**11.3.2** N 2, N 10)
C. vahlianum is found in East and North Greenland north of Scoresby Sund (Bay 1992). The species prefers soils rich in carbonates. It grows in wet sand, on constantly irrigated slopes below snow-beds and on moving-soils. Due to these ecological demands *C. vahlianum* has been collected only in the sedimentary parts of WA1, WA3, WA4 and WA5, and was not found in the mountains of WA2.

Three sites are situated above 800 m in geographically separated regions (WA1, north of Pingo Pass: 1000 m; WA3, east of Junctiondal: 1160 m, 1210 m). In addition there are four collections from the nunataks north of 74°N with altitudes between 640 m and 780 m (Schwarzenbach 1961).

Minuartia biflora ADT C2
M. biflora is a southern species (Bay 1992). It reaches the northern limit at the East coast at Skærfjorden (Böcher at al. 1978). *M. biflora* grows in snow-patches, heaths and also in fell-field.
M. biflora has been collected in WA1-WA4. It was common in WA1 around Pingo Pass. The species grows in low and middle altitudes. There are only a few collections from altitudes of 700 m ore more (WA1, east of Pingo Pass: 840 m; WA2, Sefstrøm Gletscher: 1030 m; WA3, Morænedal: 700 m; WA4, North Strindberg Land: 960 m).

Oxyria digyna ADT C2
O. digyna is a common circumgreenlandic species (Bay 1992). It grows in heaths, on herb-slopes and in fissures of rocks (Böcher et al. 1978).

This species is more or less bound to the sedimentary parts of WA1-WA5 and seems to be a calciphyte. *O. digyna* was not collected in the interior parts of the crystalline Stauning Alper. The species has been found at low and middle altitudes below 1000 m with the exception of two sites (WA1, north of Pingo Pass: 1000 m; WA3, east of Junctiondal: 1150 m).

Ranunculus affinis s.l. ADT C2 (**11.3.2** N 4, N 10, N 13)
Böcher et al. (1978) describe three forms of *R. affinis* based on cytotaxonomy. Bay (1992) shows the tricentric distribution of *R. affinis s.l* on a map. The southern limit is in Scoresby Sund. The species often grows along running melt-water below snow-beds, on wet grus or amongst mosses. It flowers early in the season.

R. affinis s.l. was collected in WA2-WA5. It was not found in WA1 though it has been reported from Fynselv in Jameson Land (Fredskild et al. 1986). The species has not been collected in the western continental belt with one exception (WA4, Krumme Langsø: 200 m).

Most of the sites were situated above 900 m. (WA2, Sefstrøm Gletscher: 960 m, 1200 m; WA3, east of Junctiondal: 970 m, 1100 m, 1200 m, Kap Weber: 970 m).

R. affinis s.l. has also been found in WA5, Ingolf Fjord: 65 m.

Ranunculus hyperboreus ADT C2 (**11.2.3** N 10; **11.4.9**)
R. hyperboreus is nearly circumgreenlandic, missing only in the coastal belt of North Greenland (Bay 1992). The species is restricted to wet places. It may be found amongst mosses, on wet sand or on clay near ponds where water percolates after snowmelt. (Böcher et al. 1978).

Due to the ecological demands the species is rare in the mountains of the western continental belt. There are three collections only (WA3, east of Junctiondal: 1150 m; WA4, Bartholin Land: 900 m; WA5: Sæfaxi Dal: 200 m).

Ranunculus sulphureus ADT C2
R. sulphureus has a northern distribution (Bay 1992) with the southern limit in East Greenland at Kap Dalton (Böcher et al. 1978). The species is found in snow-patches, on ground soaked by running water or amongst mosses.

R. sulphureus was fairly common in WA1, around Pingo Pass. Further to the north it was found only twice (WA4, Vibeke Sø: 780 m; WA5, Rivieradal: 250 m).

6.4.3 Altitude distribution type (and class) ADT C3

General comments

There are five species of this type.

Comments on the individual species

Equisetum arvense s.l. ADT C3
E. arvense s.l. is a circumgreenlandic species (Bay 1992). It grows on humid, often clayey soils and is an element of many types of vegetation.

E. arvense s.l. has been collected in all five study areas. It is a species of low

Tab. 23. Altitude distribution type (and class) ADT C3: Highest altitude collections in WA1-WA5

ADT	Species	WA1	WA2	WA3	WA4	WA5
C3	*Equisetum arvense s.l.*	760	230	650	230	75
C3	*Equisetum variegatum*	1000	*	650	200	310
C3	*Erigeron humilis*	840	1030	700	240	*
C3	*Eutrema edwardsii*	500	*	380	780	70
C3	*Juncus arcticus*	520	*	380	200	*
C3	*Koenigia islandica*	500	20	720	780	*

ADT Altitude type
WA1-WA5 Study areas
* No specimens available
Altitudes in m

and middle altitudes. The highest altitudes in WA1-WA5 are all below 800 m (WA1, north of Pingo Pass: 760 m; WA2, Schaffhauserdalen: 230 m; WA3: Moraenedal: 650 m; WA4: Vibeke Sø: 230 m; WA5, Centrum Sø: 75 m).

Equisetum variegatum ADT C3 (**11.4.7**)

E. variegatum has almost a circumgreenlandic distribution, missing only in a part of Northwest Greenland (Bay 1992). It is a calciphilous species and is found in many types of vegetation.

E. variegatum was found in WA1, WA3-WA5, but not in the Stauning Alper (WA2). The species has been collected at low and middle altitudes. There is only one high collection (WA1, north of Pingo Pass, 1000 m), where the species was found in a constantly irrigated polygon-field frequented by geese and other birds.

Erigeron humilis ADT C3 (**11.3.2** N 4; **11.4.8**)

E. humilis has a bicentric distribution in Greenland (Bay 1992). It is found in East Greenland between 63°N and c. 77°N (Böcher et al. 1978, Fredskild & Bay 1990). The species grows in snow-patches and on herb-slopes.

E. humilis has been found in WA1-WA4. It is a species of low and middle altitudes, and has been collected mostly below 900 m. However, there are three sites above this level in the Stauning Alper (WA2, Sefstrøm Gletscher: 960 m, 1010 m, 1030 m).

Eutrema edwardsii ADT C3

E. edwardsii is a bicentric northern species (Bay 1992). The species is found at the East coast of Greenland north of c. 70°N (Böcher et al. 1978). It is a calciphilous species and grows often in mires and on irrigated polygon-fields.

E. edwardsii been collected only at five sites (WA1, Pingo Pass: 500 m; WA3, south of Moraenedal: 380 m; WA4: Vibeke Sø: 780 m, WA5, Rivieradal: 70 m, Ingolf Fjord: 65 m).

Juncus arcticus ADT C3
J. arcticus is a southern species (Bay 1992) and reaches its northern limit in East Greenland at 74°24'N (Böcher et al. 1978).

J. arcticus has been collected in WA1, WA3 and WA4. The highest collections are below 550 m (WA1, Pingo Pass: 520 m; WA3: Moræenedal: 380 m; WA4, Krumme Langsø: 200 m).

Koenigia islandica ADT C3 (Fig. 27)
K. islandica is recorded from all parts of East and Northeast Greenland (Bay 1992). The species grows on wet sand and gravel, it is often found on soli-fluction soils and in snow-patches, further in cushions of moss and along running water (Böcher et al. 1978).

K. islandica was collected in WA1-WA4 at low and middle altitudes (WA1, east of Pingo Pass: 500 m; WA2, east of Linné Gletscher: 20 m; WA3, Moræne-dal: 480 m, 720 m; WA4: Vibeke Sø: 780 m).

6.5 Comments on the altitude type ADT D, D1-D3

6.5.1 Altitude distribution type (and class) ADT D1

General comments
Juncus biglumis is the only species of this group.

Comments on the individual species
Juncus biglumis ADT D1 (**11.3.2** N 1**)**
J. biglumis is a common circumgreenlandic species (Bay 1992). The species grows on humid soils with clay or grus, often at places, where the snow melts late (Böcher et al. 1978).

J. biglumis has been collected in all five study areas. However, it was rare in WA2 due to the ecological conditions in the Stauning Alper and in Nathorst

Tab. 24. Altitude distribution type (and class) ADT D1: Highest altitude collections in WA1-WA5

ADT	Species	WA1	WA2	WA3	WA4	WA5
D1	*Juncus biglumis*	1200	800	1200	1180	700

ADT Altitude type
WA1-WA5 Study areas
* No specimens available
Altitudes in m

Land. *J. biglumis* is a species of low and middle altitudes with the exception of a few regions at high altitudes (WA1, north of Pingo Pass: 1000 m, 1200 m; WA3, east of Junctiondal: 1140 m, 1150 m, upper Junctiondal: 1200 m; WA4, nunatak in North Strindberg Land: 960 m, Bartholin Land: 900 m, 1180 m; WA5, Drømmebjerg: 700 m).

6.5.2 Altitude distribution type (and class) ADT D2

General comments

The type (and class) ADT D2 includes seven species, among them one species of the genus *Eriophorum*. The highest sites of the seven species exceed only in five cases the limit of 900 m. In WA2 (Stauning Alper and in Nathorst Land) the ecological conditions are unfavourable what explains the rather low altitude limit.

Comments on the individual species

Alopecurus alpinus ADT D2 (**11.3.2** N 10)

The map published by Bay (1992) shows that *A. alpinus* is rather common and evenly distributed in North and Northeast Greenland beyond 74°N. The species is typical of places with wet sand. It prefers calcareous ground and manured soils.

South of 73°N *A. alpinus* has been found only once (WA2, Højedal: 900 m). No specimens are available from the Stauning Alper (WA2) and from the southern Werner Bjerge (WA1).

A. alpinus is rather common in low and middle altitude bands. There are a few collections from levels above 900 m (WA3, east of Junctiondal: 1000 m, 1140 m, 1150 m; WA4, nunatak in North Strindberg Land: 960 m, Bartholin Land: 960 m, 1180 m, 1200 m). The species has been collected frequently north

Tab. 25. Altitude distribution type (and class) ADT D2: Highest altitude collections in WA1-WA5

ADT	Species	WA1	WA2	WA3	WA4	WA5
D2	*Alopecurus alpinus*	*	900	1150	1200	700
D2	*Arctagrostis latifolia*	1000	900	720	780	310
D2	*Carex atrofusca*	480	1310	1150	780	40
D2	*Eriophorum triste*	1000	900	1140	900	40
D2	*Juncus triglumis*	690	690	720	900	40
D2	*Kobresia simpliciuscula*	690	*	720	900	75
D2	*Saxifraga aizoides*	1000	70	650	640	75

ADT Altitude type
WA1-WA5 Study areas
* No specimens available
Altitudes in m

of 80°N in Kronprins Christian Land, where there are good conditions on the large areas with active polygon-fields. It reaches rather high altitudes (WA5, Drømmebjerg: 700 m).

Arctagrostis latifolia ADT D2 (**11.3.2** N 2, N 9)
The northern species *A. latifolia* is distributed in East and North Greenland and occurs in an isolated area in West Greenland (Bay 1992). It reaches the southern limit at the eastern coast in Jameson Land, 70°20'N (Böcher et al. 1978).

A. latifolia has been collected in all five study areas WA1-WA5. The species is rather common in areas with sediments. It often grows on terraces, in depressions and at the foot of gentle slopes. As the species depends on a long-lasting supply of water during the growth period it is often found in mires, in wet heaths and in ponds. *A. latifolia* has been collected mostly at low and middle altitudes. However, there are a few exceptionally high sites (WA1, north of Pingo Pass: 1000 m; WA2, Højedal: 900 m; WA3, Moræenedal: 720 m; WA4, Vibeke Sø: 780 m; WA5, Centrum Sø: 310 m).

Carex atrofusca ADT D2 (**11.3.2** N 7, N 10; **11.4.8)**
C. atrofusca has a multicentric distribution in Greenland (Bay 1992). Böcher et al. (1978) state that the species is found in East Greenland between Jameson Land (70°26'N) and Mørkefjord (77°55'N). Bay & Fredskild (1990 b) mention a record from 78°33'N. *C. atrofusca* is a calciphilous species and grows in mires.

The species has been collected in WA1-WA4 and exceptionally also in WA5: Centrum Sø, 80°07'N, 22°16'W, 35 m, and 80°07'N, 22°23'W, 40 m (Bay & Fredskild 1994).

C. atrofusca is a plant of low and medium altitude with two remarkable exceptions (WA2, Skjoldungebræ: 1310 m; WA3, Junctiondal: 1150 m).

Eriophorum triste ADT D2 (**11.3.2** N 2, N 9, N 10. **11.4.7**)
E. triste has a bicentric distribution in Greenland (Bay 1992). The southern limit is at c. 70°N (Böcher et al. 1978). The species grows on humid soils, in mires and on wet flats along lakes and ponds.

The species has been collected in all five study areas. It has been found in general below 700 m with a few exceptions. At the highest site in WA1 (North of Pingo Pass: 1000 m) the species grows in a permanently wet polygon-field together with *Cochlearia groenlandica, Equisetum variegatum* and *Luzula arctica*. This site is thought to be a site of pioneer-plants. In WA2 and WA3 there are four collections at high altitudes. (WA2, Højedal: 900 m; WA3, east of Junctiondal: 970 m, 1150 m, Moræenedal: 720 m). Two sites in the nunataks of 74°N may have been reached in recent times (WA4, Bartholin Land: 900 m, C. H. Ostenfeld Nunatak: 700 m). In WA5 the species was collected only at levels below 50 m.

Juncus triglumis ADT D2
J. triglumis is nearly circumgreenlandic missing only in part of Northwest Greenland (Bay 1992). The species is calciphilous and grows on wet soils in mosses and on wet flats near lakes (Böcher et al. 1978).

J. triglumis is quite common in East and Northeast Greenland. The basiphilous species has been collected in all five study areas. *J. triglumis* was often found in the western continental belt between the Schuchert Dal (WA1) and the nunataks north of 74°N (WA4).

The species has been observed only from the three lowermost altitude bands.The altitudes of highest sites in WA1-WA4 range between 690 m and 900 m (WA1, Biskop Alf Dal: 690 m; WA2, Tærskeldal: 690 m; WA3, Moræne-dal: 720 m; WA4, Bartholin Land: 900 m). *J. triglumis* was not found on the plateau above 1000 m in WA3, dominated by calcareous and dolomitic sedi-ments.

Kobresia simpliciuscula ADT D2
K. simpliciuscula has a tricentric distribution in Greenland. At the East coast the species has been recorded north of Scoresby Sund (Bay 1992). The species is restricted to mires and heath. It prefers calcareous soils (Böcher at al. 1978).

The species was found in WA1, WA3-WA5. It has not been collected in WA2, but there is no doubt that the species occurs in the alluvial parts of that study area. *K. simpliciuscula* is a species of lower and middle altitude (Maxima: WA1, Biskop Alf Dal: 690 m; WA3, Moraenedal: 720 m; WA4, Bartholin Land: 900 m; WA5, Centrum Sø: 75 m).

Saxifraga aizoides ADT D2 (**11.3.2** N 2; **11.4.6, 11.4.7**)
The species has a southern distribution, the northernmost record is from Centrum Sø in Kronprins Christian Land (Böcher et al. 1978). *S. aizoides* grows on wet soils, in mires and along running water.

S. aizoides has been collected in all study areas WA1-WA5. It prefers calcareous and dolomitic soils and is rather common in the sedimentary parts of the study areas.

In WA1, southern Werner Bjerge, the species was found in low and middle altitude below 800 m. There is one exception (WA1, north of Pingo Pass: 1000 m), where some lowland species (e.g. *Cochlearia groenlandica, Silene acaulis, Eriophorum scheuchzeri*) grow in a permanently wet polygon-field. There are a few high collections from two other study areas (WA3, Moraenedal: 650 m; WA4, C. H. Ostenfeld Nunatak: 640 m).

6.5.3 Altitude distribution type (and class) ADT D3

General comment

There are six species in the type (and class) ADT D3. They are all restricted to low altitude, and they are missing in WA5.

Tab. 26. Altitude distribution type (and class) ADT D3: Highest altitude collections in WA1 - WA5

ADT	Species	WA1	WA2	WA3	WA4	WA5
D3	*Carex saxatilis*	690	800	680	230	*
D3	*Eriophorum callitrix*	520	40	25	700	40
D3	*Eriophorum scheuchzeri*	610	380	1150	900	700
D3	*Juncus castaneus*	690	690	720	240	40
D3	*Saxifraga foliolosa*	690	85	*	*	*
D3	*Saxifraga nathorstii*	240	*	480	900	*

ADT Altitude type
WA1-WA5 Study areas
* No specimens available
Altitudes in m

Comments on the individual species

Carex saxatilis ADT D3

C. saxatilis is a southern species with a few isolated observations from North and Northwest Greenland (Bay 1992).

C. saxatilis has been collected in WA1-WA4. Ecologically restricted to mires and banks of ponds or lakes it has been found only in the three lowest altitude zones. The highest sites in WA1-WA4 are given in the following list (WA1, Biskop Alf Dal: 690 m; WA2, Højedal: 800 m; WA3, Moraenedal: 700 m; WA4, Vibeke Sø: 230 m). This species has not been collected in WA5, but was found 1995 in North Peary Land (Bay 1992).

Eriophorum callitrix ADT D3

E. callitrix is a tricentric species with the main area in East and Northeast Greenland (Bay 1992). The southernmost record is from Gåseland, 70°15'N (Böcher et al. 1978). It grows in mires.

This species has been found in all five study areas at low and middle altitude up to 700 m. The most remarkable collections are from WA4 and from WA5 (WA4, C. H. Ostenfeld Nunatak: 415 m, 700 m; WA5, Centrum Sø: 40 m, Fyn Sø: 10 m).

Eriophorum scheuchzeri ADT D3 (**11.3.2** N 10, N 17; **11.4.7**)

E. scheuchzeri is a circumgreenlandic species (Bay 1992). It grows in mires, in wet heaths, along running water and on flats near ponds and lake.

E. scheuchzeri has been collected in WA1-WA5. It has been found at altitudes below 700 m with a few exceptions (WA3, east of Junctiondal: 1150 m, Moraenedal: 720 m; WA4, Bartholin Land: 900 m, C. H. Ostenfeld Nunatak: 700 m; WA5, Drømmebjerg: 700 m).

Juncus castaneus ADT D3
J. castaneus has a tricentric distribution. The species is found in East Greenland between Angmagssalik and Peary Land (Bay 1992). It grows on wet, but not acidic soils, in mosses and mires (Böcher et al. 1978).

J. castaneus has been collected in all five study areas WA1-WA5. It was found in low and middle altitude (maxima: WA1, Biskop Alf Dal: 690 m; WA2, Tærskeldal: 690 m; WA3, Morænedal: 720 m; WA4, Vibeke Sø: 240 m; WA5, Centrum Sø: 40 m).

Saxifraga foliolosa ADT D3
The species *S. foliolosa* is found in most parts of Greenland except the south. (Bay 1992). It grows in snow-patches, near springs, along running water and also in mires (Böcher et al. 1978).

S. foliolosa has only been collected in WA1 and WA2 restricted to the lowest three altitude bands. The species was quite common in WA1, southern Werner Bjerge, with the highest collection from Biskop Alf Dal: 690 m. Two specimens have been found in WA2, Segelsällskapet Fjord: 85 m, Forsblad Fjord, Caledonia Ø: 15 m.

Saxifraga nathorstii ADT D3 (**11.4.6**)
S. nathorstii is a monocentric species (Bay 1992). It is found in East and Northeast Greenland between Scoresby Sund and Hochstetter Forland (Böcher et al. 1978). The species has similar ecological demands as *S. aizoides* and grows along running water, in snow-patches, in wet heaths as well as in mires. The species prefers calcareous soils.

There are three geographically separated localities in WA1, Schuchert Dal: 100 m, Kong Oscar Fjord, near Skjoldungebræ: 240 m, Kap Peterséns: 110 m.

The species was found five times in North Andrée Land (WA3), with the highest site in Morænedal (480 m). *S. nathorstii* was also found at six sites in the nunataks north of 74°N (WA4, Krumme Langsø: 200 m, Vibeke Sø: 780 m, Bartholin Land: 900 m, C. H. Ostenfeld Nunatak: 415 m, 640 m, 700 m).

6.6 Comments on the altitude distribution type ADT E, E3

6.6.1 Altitude distribution type (and classe) ADT E3

General comments

The type (and classes) ADT E3 include the fresh-water plants *Ranunculus confervoides, Hippuris vulgaris* and *Potamogeton filiformis*. All three species have been collected only occasionally.

Tab. 27. Altitude distribution type (and class) ADT E3: Highest altitude collections in WA1-WA4

ADT	Species	WA1	WA2	WA3	WA4	WA5
E3	*Ranunculus confervoides*	*	*	680	200	*
E3	*Hippuris vulgaris*	*	*	240	380	*
E3	*Potamogeton filiformis*	*	*	*	230	*

ADT Altitude type
WA1-WA5 Study areas
* No specimens available
Altitudes in m

Comments on the individual species

Ranunculus confervoides ADT E3

R. confervoides has a multicentric distribution in Greenland (Bay 1992). It is rare in North and Northeast Greenland. The species has been collected in WA3 and WA4. There are only two collections (WA3, Moranedal: 680 m; WA4, Krumme Langsø: 200 m).

Hippuris vulgaris ADT E3

H. vulgaris has a multicentric distribution in Greenland (Bay 1992). It grows in ponds and at lakes with a high content of nutrients (Böcher et al. 1978). *H. vulgaris* has been found in WA3, Moranedal: 240 m and in WA4, Vibeke Sø: 380 m.

Potamogeton filiformis ADT E3

P. filiformis is a southern species (Bay 1992). It reaches the northern limit at 76°52'N (Fredskild & Bay 1990). The species has been collected only once (WA4, Vibeke Sø: 74°07'N, 23°32'W, 230 m).

6.7 Comments on the altitude distribution type ADT F, F3

6.7.1 Altitude distribution type (and class) ADT F3

General comments

There is no list of the highest sites because the four species *Carex glareosa, C. ursina, Honckenya peploides* and *Stellaria humifusa* are coastal plants.

Comments on the individual species

Carex glareosa ADT F3

C. glareosa is found south of 78°N in West Greenland, south of 75°N at the East coast (Böcher et al. 1978). This species grows near the shore. The species has only been found in WA2, Caledonia Ø and at the western end of Forsblad Fjord.

Carex ursina ADT F3
C. ursina is a tricentric species (Bay 1992) reaching c. 77°N in East Greenland (Fredskild & Bay 1990). This species grows in ‚meadows' at the sea-shore. *C. ursina* has been collected only once at the western end of Forsblad Fjord.

Honckenya peploides ADT F3
H. peploides is a southern species with an isolated area in Northwest Greenland (Bay 1992). The northernmost locality in East Greenland (Fredskild & Bay 1990) is Sedimentkløft in Germania Land (78°34'N, 21°30'W). The species grows in wet sand near the sea-shore (Böcher et al. 1978).

This species has been found in WA2 and WA3. There are three collections from WA2 along Forsblad Fjord. Two more specimens were collected in WA3, Benjamin Dal and Morænedal.

Stellaria humifusa ADT F3
S. humifusa has been classified by Bay (1992) as a southern species, reaching c. 79°30'N on the northern coast of Greenland. This species is found at the sea-shore. *S. humifusa* has been collected only in WA2, Forsblad Fjord, Caledonia Ø.

6.8 Comments on the altitude distribution type ADT G, G3

6.8.1 Altitude distribution type (and class) ADT G3

General comments

These 28 rare species are found in WA1-WA3. Most of the species reach their northern limit south of 75°N. They have altitude limits below 900 m.

Comments on the individual species

Arabis alpina ADT G3 (**11.3.2** N 3)
A. alpina is a southern species and reaches the northern limit just north of 74°N (Bay 1992). It grows at sites with bushes of *Salix sp.*, on herb-slopes and in snow-patches (Böcher et al. 1978).

A. alpina follows the sedimentary area east of Schuchert Dal from Jameson Land through the eastern Werner Bjerge to Skeldal and Syltoppene (WA1, Pingo Pass: 620 m, near Biskop Alf Gletscher: 770 m; Syltoppene: 70 m, 620 m).

Arabis holboellii ADT G3 (**11.3.2** N 5; **11.4.8;** Fig. 33)
A. holboellii is a southern species, found at the East coast of Greenland only be-

Tab. 28. Altitude distribution type (and class) ADT G3 Highest altitude collections WA1-WA4

ADT	Species	WA1	WA2	WA3	WA4	WA5
G3	*Arabis alpina*	770	*	*	*	*
G3	*Arabis holboellii*	*	360	*	*	*
G3	*Arenaria humifusa*	240	*	*	*	*
G3	*Botrychium lunaria*	770	*	*	*	*
G3	*Carex bicolor*	75	*	*	*	*
G3	*Carex lachenalii*	620	*	*	230	*
G3	*Carex marina s.l.*	550	*	*	230	*
G3	*Carex microglochin*	80	*	*	230	*
G3	*Carex parallela*	*	100	25	*	*
G3	*Carex rariflora*	*	230	*	230	*
G3	*Draba aurea*	*	360	*	*	*
G3	*Draba crassifolia*	*	*	2	*	*
G3	*Gentiana detonsa*	*	*	40	*	*
G3	*Gentiana tenella*	770	*	380	*	*
G3	*Harrimanella hypnoides*	90	1030	*	*	*
G3	*Juncus trifidus*	80	*	*	*	*
G3	*Lycopodium annotinum*		270	*	*	*
G3	*Minuartia stricta*	510	*	400	*	*
G3	*Pinguicula vulgaris*	100	10	*	*	*
G3	*Potentilla crantzii*	770	810	*	*	*
G3	*Ranunculus nivalis*	620	20	*	*	*
G3	*Rhodiola rosea*	*	40	85	*	*
G3	*Sagina caespitosa*	*	*	480	*	*
G3	*Sibbaldia procumbens*	770	1030	*	*	*
G3	*Taraxacum brachyceras*	770	1030	*	*	*
G3	*Thalictrum alpinum*	770	*	*	*	*
G3	*Triglochin palustre*	100	*	60	*	*
G3	*Viscaria alpina*	*	810	*	*	*

ADT Altitude type
WA1-WA5 Study areas
* No specimens available
Altitudes in m

tween 70°15'N and 72°40'N (Böcher et al. 1978). The species grows on dry open sites, in willow scrub and on dry slopes.

It was a real surprise to find *A. holboellii* near Vikingebræ in the Stauning Alper at an altitude of 360 m (72°12'N, 25°22'W). A considerable number of vigorous individuals were growing along a dry sunny slope. They had plenty of pods with many ripe seeds at August 14, 1954.

Arenaria humifusa ADT G3

A. humifusa has a tricentric distribution in Greenland (Bay 1992). I grows on wet sand or grus and in fens, preferring calcareous soils.

The only specimens have been collected in WA1 (26/07/1951) near the end of Skjoldungebræ c. 5 km from the coast at an altitude of 240 m (72°22'N,

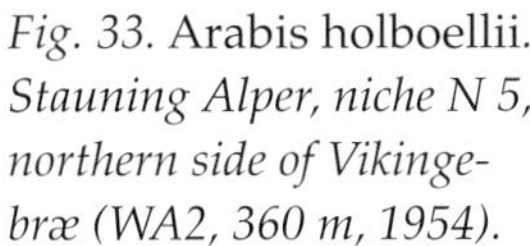

Fig. 33. Arabis holboellii. *Stauning Alper, niche N 5, northern side of Vikingebræ (WA2, 360 m, 1954).*

24°36'W). The plants were growing in a small mire. Not recognising the species – at that time not known from this part of the East coast – the plant was erroneously determined by the author as *Minuartia stricta*. When Dr G. Halliday revised the writer's collection of 1951 he corrected the error (Böcher 1963: 30). The same author writes that Laegaard has found *A. humifusa* later at three places in Scoresby Sund and that Dr G. Halliday mentions six observations from the area around Kap Peterséns in Kong Oscar Fjord. Since 1961 *A. humifusa* has been found in other places too. The northernmost record is from Hochstetter Forland 75°19'N (Halliday 1981, Bay 1992).

Botrychium lunaria ADT G3 (**11.3.2** N 3; **11.4.8**)
B. lunaria is found in West and East Greenland with the northern limit at c. 73°N. It grows on herb-slopes covered by snow in winter-time and has been recorded from warm springs (Böcher et al. 1978). *B. lunaria* has been collected only once (WA1, Biskop Alf Dal: 770 m).

Carex bicolor ADT G3
C. bicolor is a bicentric species. It is found in East Greenland between Skjoldungen and 73°37'N (Böcher et al. 1978). The species prefers calcareous soil and grows on wet sand and clay as well as on strand-terraces and in mires. The species has been found only once (WA1, Bersærkerbræ, 72°13'N, 24°19'W, 90 m). *C. bicolor* grows there on a dry soil rich in mineral salts. *C. bicolor* demands similar ecological conditions as e.g. *Lesquerella arctica, Braya humilis* or *Armeria scabra* ssp. *sibirica*.

Carex lachenalii ADT G3
C. lachenalii has been found northward to 74°37'N (Böcher et al. 1978). It grows in snow-patches and on herb-slopes. The species has been collected in WA1 and WA4 (WA1, Pingo Pass: 500 m, 620 m, Bersærkerbræ: 90 m; WA4, Krumme Langsø: 200 m; Vibeke Sø: 230 m).

Carex marina s.l. ADT G3
C. marina s.l. has a multicentric distribution in Greenland, but is not present in the southern parts (Bay 1992). Fredskild and Bay (1990) have found the species at several places between 75°N and 78°N with the northernmost record from Germania Land (77°35'N). Schwarzenbach has collected *Carex marina s.l.* twice in North Peary Land, shifting the northern limit to a considerable extent (Fredskild 1996 b):

- 02.07.1995, Frigg Fjord, Grønnemarken, west of Nysne Gletscher: 83°16'N, 34°13'W, 320 m.
- 07.07.1995, Frigg Fjord, Grønnemarken, western side of the main river: 83°12'N, 34°12'W, 210 m.

Carex marina s.l. grows in mires with mosses and along lakes. This species was collected in two sites only (WA1, Pingo Pass: 550 m; WA4, Vibeke Sø: 230 m).

Carex microglochin ADT G3
C. microglochin is a southern species (Bay 1992) with the northern limit in East Greenland at 74°30'N (Böcher et al. 1978).The species grows in mires and on wet flats of clay. This rare species was found twice (WA1, Schuchert Dal: 80 m; WA4, Vibeke Sø: 230 m). According to Hess et al. (1967) the arctic/alpine *C. microglochin* is rare in the Swiss Alps too and has been found only at a few isolated places. It seems that the species there is an interglacial relict.

Carex parallela ADT G3
C. parallela has a monocentric distribution and is only found at the eastern coast of Greenland (70°15' – 74°22'N, preferring the inland (Böcher et al. 1978). The species grows in mires or on wet clayey flats.

C. parallela has been collected at five sites in WA2-WA3, in the middle and inner parts of the fjords. The species was only found at low altitudes not exceeding 100 m.

Carex rariflora ADT G3
C. rariflora is a southern species and has the northern limit in East Greenland at 74°30'N (Böcher et al. 1978). It grows in moss, in fens and in meadows near the shore.

C. rariflora has been collected in two areas only (WA2, Linné Gletscher: 5 m, 60 m, Schaffhauserdalen: 160 m, 230 m, WA4, Vibeke Sø: 230 m).

Draba aurea ADT G3 (**11.3.2** N 5; **11.4.8;** Fig. 34)
D. aurea is a southern species with the northern limit at 67°N in East Greenland. However, there are two isolated sites further north: Gåseland (70°N) and the Stauning Alper (72°N), as is stated by Böcher et al. 1978). *D. aurea* has been found at two sites (WA2, north of Vikingebræ: 150 m, 260 m).

Draba crassifolia ADT G3
D. crassifolia has a bicentric distribution (Bay 1992). It is found in East Greenland between Angmagssalik and Clavering Ø (74°N). The species grows in snow-beds and also on manured soils (Böcher et al. 1978). The species has been collected only in WA1, Pingo Pass: 620 m.

Gentiana detonsa ADT G3
G. detonsa is a southern species. The species grows on alkaline, often saline soils drying out during the summer (Böcher et al. 1978). The only specimen of *G. detonsa* has been collected in WA3 (Junctiondal: 40 m). Several plants were in flower on a steep sunny slope between rocks. This rare species is known from several places on the northern side of the innermost Scoresby Sund.

Gentiana tenella ADT B3 (**11.3.2** N 3; **11.4.8**)
G. tenella has a bicentric distribution in West and East Greenland (Bay 1992).

Fig. 34. Draba aurea. *Stauning Alper, niche N 5, northern side of Vikingebræ (WA2, 260 m, 1954).*

In East Greenland the species is known from the area between Jameson Land and Germania Land (Böcher et al. 1978). It grows on dry alkaline soils which dry out during the summer.

G. tenella was collected in WA1 and WA3. It follows the belt of sediments east of the Stauning Alper. The species has been found four times in WA1 around Pingo Pass, the highest site is west of Biskop Alf Dal: 770 m. *G. tenella* was also collected in WA3, Morænedal: 380 m.

Harrimanella hypnoides ADT G3 (**11.3.2** N 4; **11.4.8**)
H. hypnoides is a southern species with the northern limit in East Greenland at c. 75°N (Bay 1992). It grows in snow-patches (Böcher et al. 1978). *H. hypnoides* was found only twice (WA1, Bersærkerbræ: 90 m; WA2, Sefstrøm Gletscher: 1030 m).

Juncus trifidus ADT G3
J. trifidus is a southern species. Böcher et al. (1978) set the northern limit in East Greenland at c. 72°30'N. Bay (1992) and maps a record at c. 75°N. *J. trifidus* has been collected only once (WA1, Bersærkerbræ: 80 m), but it is expected that there are more sites in the lowest altitude bands around the Stauning Alper.

Lycopodium annotinum ADT G3
L. annotinum is a southern species typical for heath and scrub covered by snow in the winter. In East Greenland the northern limit is 72°30'N (Böcher et al. 1978). *L. annotinum* was found at two sites (WA2, Schaffhauserdalen: 270 m, Forsblad Fjord, Rondenæs: 5 m).

Minuartia stricta ADT G3
M. stricta is a tricentric species (Bay 1992). It is found in East Greenland between Scoresby Sund and Daneborg. The species grows on grus, on sand or in heaths (Böcher et al. 1978).

M. stricta has been found three times in both subareas of WA1 (20-250 m) and in WA3, Morænedal: 400 m.

Pinguicula vulgaris ADT G3
The southern species *P. vulgaris* reaches c. 73°N in East Greenland (Böcher et al. 1978). The species has been found three times (WA1, Schuchert Dal: 80 m, 100 m; WA2: western end of Forsblad Fjord: 10 m).

Potentilla crantzii ADT G3 (**11.3.2** N 3, N 5; **11.4.8**)
P. crantzii is a southern species reaching c. 74°N at the East coast (Bay 1992). The species is a characteristic plant of herb-slopes, scrub and heaths rich in species.

Fig. 35. Rhodiola rosea. *Forsblad Fjord (WA2, 10 m, 1954).*

P. crantzii has been found in WA1 and WA2. It was collected at eight sites in WA1, Pingo Pass, at middle altitudes. The highest site is behind the old lateral moraine of Biskop Alf Gletscher: 770 m. The species has also been found at a remote site in WA2, north of Vikingebræ: 810 m. .

Ranunculus nivalis ADT G3
R. nivalis is found along the East coast north of 69°30'N (Böcher et al. 1978). The species grows below snow-beds, along running water or in mosses.

R. nivalis has been found four times in the sedimentary parts of WA1 and WA2 at low and middle altitude (WA1, Pingo Pass: 510 m, 610 m, 620 m; WA2, east of Linné Gletscher: 20 m).

Rhodiola rosea ADT G3 (Fig. 35)
R. rosea is a southern species (Bay 1992) and reaches c. 75°N at the East coast. It grows in fissures of rocks, on herb-slopes and in heaths (Böcher et al. 1978). *R. rosea* has only been collected in two areas (WA2, Segelsällskapet Fjord: 85 m, Forsblad Fjord: 10 m; WA3, Junctiondal: 40 m).

Sagina caespitosa ADT G3
S. caespitosa is a bicentric species (Bay 1992). It is found in East Greenland be-

tween Hold with Hope and Mørkefjord (Böcher et al. 1978). The species is rare. It grows on sand and grus. *S. caespitosa* has been collected only once (WA3, Moraenedal: 480 m).

Sibbaldia procumbens ADT G3 (**11.3.2** N 3, N 4, N 5; **11.4.8**)
S. procumbens is a southern species with the northern limit in East Greenland at c. 74°20′N (Bay 1992). It is found in herb-slopes and in early melting snow-patches (Böcher et al. 1978).

This species was collected at several sites of unexpectedly high altitude. (WA1, Biskop Alf Dal: 770 m; WA2, Sefstrøm Gletscher: 1010 m, 1030 m, north of Vikingebræ: 810 m).

Taraxacum brachyceras ADT G3 (**11.3.2** N 3, N 4, N 5; **11.4.8**)
This monocentric species is found in East Greenland only. Böcher et al. (1978) set the southern limit to 68°10′N, the northern limit to 74°10′N. It grows on herb-slopes, in early-melting snow-patches as well as on sunny slopes.

T. brachyceras was collected at five sites in WA1 and WA2 (WA1, east of Pingo Dal: 650 m, Biskop Alf Dal: 770 m; WA2, Sefstrøm Gletscher: 1030 m, north of Vikingebræ: 360 m, 810 m).

Thalictrum alpinum ADT G3 (**11.3.2** N 3; **11.4.8**)
T. alpinum has a southern distribution with the northern limit at 74°10′N (Böcher et al. 1978). The species grows on herb-slopes, in scrub and with grasses at low altitude.

T. alpinum was collected at three sites (WA1, south of Pingo Dal: 650 m, Biskop Alf Dal: 770 m, Bersærkerbræ: 90 m). It seems that *T. alpinum* has reached a northern outpost in the southern Werner Bjerge (11.4.8).

Triglochin palustre ADT G3
T. palustre has a southern distribution. The northern limit in East Greenland is about 74°N (Böcher et al. 1978). The species grows on clayey terraces, near the shore, in the alluvial parts of rivers and on soils with a high content of mineral salts. The species has been found in two areas (WA1, Schuchert Dal: 100 m; WA3, Junctiondal: 60 m, Grejsdal: 30 m).

Viscaria alpina ADT G3 (**11.3.2** N 5; **11.4.8;** Fig. 36)
V. alpina is a southern species, reaching its northern limit in East Greenland at 73°10′N (Böcher et al. 1978). It grows in scrub, on herb-slopes and in fissures of rocks. The species was collected in WA2, north of Vikingebræ: 150 m, 810 m, southern side of Segelsällskapet Fjord: 85 m).

Fig. 36. Viscaria alpina. *Stauning Alper, niche N 5, northern side of Vikingebræ (WA2, 810 m, 1954).*

6.9 Comments on the altitude distribution type ADT H, H1-2

6.9.1 Altitude distribution type (and class) ADT H1

General comments

One of the four species has been collected in WA4 and WA5; the other three species only in WA5. The altitudes of the highest site is at 720 m in WA4, and 810 m in WA5.

Comments on the individual species

Braya thorild-wulffii ADT H1

The northern *B. thorild-wulffii* has only been collected in Kronprins Christian

Tab. 29. Altitude distribution type (and class) ADT H1: Highest altitude collections in WA1-WA4

ADT	Species	WA1	WA2	WA3	WA4	WA5
H1	*Braya thorild-wulffii*	*	*	*	*	650
H1	*Cerastium regelii* ssp. *caespitosum*	*	*	*	*	700
H1	*Epilobium arcticum*	*	*	*	*	650
H1	*Puccinellia angustata*	*	*	*	720	810

ADT Altitude type
WA1-WA5 Study areas
* No specimens available
Altitudes in m

Land. The species was found frequently in the whole study area WA5, reaching an altitude of 650 m (Drømmebjerg).

Cerastium regelii ssp. *caespitosum* ADT H1
C. regelii ssp. *caespitosum* is a monocentric species found in North and Northeast Greenland (Bay 1992). It grows in snow-patches and in active polygonfields. The bright-green cushions – mostly without flowers – often "swim" on the surface of creeping soils kept upright by the stabilising tap-root.

C. regelii ssp. *caespitosum* is widely distributed in Kronprins Christian Land. The species was common on the plateaus north of Centrum Sø reaching 700 m in the Drømmebjerg area.

Epilobium arcticum ADT H1
E. arcticum is a bicentric species. The species is rare in East and Northeast Greenland (Bay 1992). It grows in fens with mosses, on wet clay and on flats near lakes. The species prefers calcareous soils (Böcher et al. 1978). *E. arcticum* was found only at two sites (WA5, Drømmebjerg: 650 m, Centrum Sø: 35 m.

Puccinellia angustata ADT H1
The northern species *P. angustata* is found southwards to Scoresby Sund. It grows on sand and clay, as well as on moraines (Böcher et al. 1978).

P. angustata was found at four sites in WA4 on C. H. Ostenfeld Nunatak: 415 m, 640 m, 700 m, 720 m). It is possible that these specimens belong to an isolated population. In WA5 *P. angustata* has been found in three sub-areas. It was rather common on the old marine terraces along the coast of Ingolf Fjord (Maxima: WA5, Drømmebjerg: 700 m, north of Ingolf Fjord: 330 m, Hekla Sund: 810 m).

6.9.2 Altitude distribution type (and class) ADT H2

General comments

All nine species have been collected in WA5. In addition, *Pleuropogon sabinei*, *Saxifraga platysepala* and *Stellaria longipes s.l.* have been found in WA4. All sites are below 500 m.

Comments on the individual species

Carex stans s.str. ADT H2
C. stans s.str. is a northern species and reaches the southern limit at the East coast at c. 74°20′N (Bay 1992). The species grows in mires. Two specimens of

Tab. 30. Altitude distribution type (and class) ADT H2: Highest altitude collections in WA1-WA4

ADT	Species	WA1	WA2	WA3	WA4	WA5
H2	*Carex stans s.str.*	*	*	**	*	35
H2	*Deschampsia brevifolia*	*	*	*	*	70
H2	*Elymus hyperarcticus*	*	*	*	*	35
H2	*Pleuropogon sabinei*	*	*	*	200	35
H2	*Puccinellia bruggemanni*	*	*	*	*	600
H2	*Saxifraga platysepala*	*	*	*	780	500
H2	*Stellaria longipes s.l.*	*	*	*	780	770
H2	*Taraxaxum arctogenum*	*	*	*	*	470
H2	*Taraxacum pumilum*	*	*	*	*	200

ADT Altitude type
WA1-WA5 Study areas
* No specimens available
Altitudes in m

C. stans s.str. have been collected (WA5, west of Centrum Sø: 35 m, Centrum Sø: 35 m).

Deschampsia brevifolia ADT H2
D. brevifolia is a northern species with the southern limit in East Greenland near Kong Oscar Fjord. The species grows on open flats with sand or stones and is often found along rivers (Böcher et al. 1978, Bay 1992). *D. brevifolia* was collected four times (WA5, Centrum Sø: 30 m, Rivieradal: 70 m, Ingolf Fjord: 55 m, 65 m).

Elymus hyperarcticus ADT H2
E. hyperarcticus has a bicentric distribution (Bay 1992). The species occurs in East and Northeast Greenland between Ymer Ø (72°10'N) and Peary Land (Böcher et al. 1978). However, *E. hyperarcticus* is rather rare south of 78°N. *E. hyperarcticus* has been found in Kronprins Christian Land around Centrum Sø: 35 m)

Pleuropogon sabinei ADT H2
P. sabinei is a northern species found as far south as Jameson Land (Böcher et al. 1978). The species grows in mires and in ponds. *P. sabinei* has collected only twice (WA4, Krumme Langsø: 200 m; WA5, Centrum Sø: 35 m).

Puccinellia bruggemanni ADT H2
Bay (1993) discusses the taxon *P. bruggemanni* (not included in Böcher et al. 1978) and its distribution in Greenland. His map shows that the species is found in North and Northeast Greenland with the southernmost record at c. 75°N.

P. bruggemanni has been collected only in WA5. The plant was growing on solifluction soil (WA5, Drømmebjerg: 600 m).

Saxifraga platysepala ADT H2
S. platysepala has a northern distribution (Bay 1992). The southern limit at the East coast is at Kejser Franz Joseph Fjord, 73°N (Böcher et al. 1978). The species is found on open wet ground. It is a calciphilous species.

S. platysepala was collected four times (WA4, Vibeke Sø: 780 m, WA5, Ulvebjerg: 500 m, Fyn Sø: 30 m, Ingolf Fjord: 65 m).

Stellaria longipes s.l. ADT H2
Böcher et al. (1978) distinguish six taxa of lower rank. Bay (1992) puts them together as *S. longipes s.l.*, mapped as a circumgreenlandic species. It is found in heaths and with mosses. Some of the small-species are calciphytes, others prefer acidic soils (Böcher 1978).

S. longipes s.l. has been collected from two sites in the nunataks north of 74°N (WA4, Vibeke Sø: 780 m, C. H. Ostenfeld Nunatak: 720 m. The species was common in WA5 (Maxima: Drømmebjerg: 770 m, Ingolf Fjord: 65 m, Fyn Sø: 15 m).

Taraxacum arctogenum ADT H2 (**11.3.2** N 18)
T. arctogenum is a bicentric species found in North and Northwest Greenland (Bay 1992). It grows on clayey slopes and terraces. It is also found near lakes and at the sea-shore. The species has been collected four times in Kronprins Christian Land (WA5: Centrum Sø, 30 m, 35 m, north of Ingolf Fjord: 65 m, 470 m).

Taraxacum pumilum ADT H2
T. pumilum is found in North and Northeast Greenland (Bay 1992). The southern limit is at Shannon Ø, 75°11'N, 18°15'W (Fredskild & Bay 1989). It grows on clay, and on dry soils and grus. *T. pumilum* has been collected twice (WA5, Centrum Sø: 30 m, Sæfaxi Dal: 200 m).

6.10 Comments on the altitude distribution type ADT X

General comments

The subtype ADT X includes all species with aberrant patterns of altitude distribution. The 32 species are arranged in alphabetic order.

Comments to the individual species

Antennaria canescens ADT X (**11.3.2** N 4, N 6, N 7, N 8, N 9, N 11; **11.4.6**, **11.4.8**)
A. canescens is a southern species with the northern limit at c. 74°20'N in East

Tab. 31. Altitude distribution type ADT X Highest altitude collections in WA1-WA4

ADT	Species	WA1	WA2	WA3	WA4	WA5
X	*Antennaria canescens*	770	1335	720	*	*
X	*Antennaria porsildii*	670	1010	*	*	*
X	*Arenaria pseudofrigida*	1100	*	1250	450	*
X	*Armeria scabra* ssp. *sibirica*	100	900	400	450	80
X	*Braya linearis*	*	*	1170	200	*
X	*Carex bigelowii s.l.*	1000	850	1080	450	*
X	*Carex glacialis*	*	1300	560	*	670
X	*Carex norvegica*	210	1120	650	*	*
X	*Cochlearia groenlandica*	1000	*	*	*	250
X	*Draba adamsii s.l.*	510	*	1285	960	600
X	*Draba alpina s.l.*	1200	20	1180	450	600
X	*Draba arctogena*	*	*	1150	1370	*
X	*Draba gredinii*	1200	*	*	960	*
X	*Empetrum nigrum* ssp. *hermaphroditum*	340	960	30	*	*
X	*Erigeron compositus*	*	1150	1150	*	*
X	*Festuca hyperborea*	1200	*	150	230	670
X	*Festuca vivipara*	1100	20	1150	450	*
X	*Huperzia selago*	510	1240	440	*	*
X	*Luzula spicata*	770	1660	650	*	*
X	*Minuartia rossii*	*	*	650	*	770
X	*Phippsia algida s.l.*	620	800	1285	700	*
X	*Poa alpina s.l.*	840	1330	1140	900	*
X	*Poa pratensis s.l.*	520	710	1285	450	70
X	*Potentilla hyparctica*	*	810	1350	1270	670
X	*Pyrola grandiflora*	540	960	970	540	*
X	*Ranunculus glacialis*	*	*	1100	1350	*
X	*Ranunculus pygmaeus*	620	1200	650	*	*
X	*Sagina intermedia*	690	10	*	780	70
X	*Salix herbacea*	*	1030	420	*	*
X	*Saxifraga hieracifolia*	520	*	*	*	*
X	*Saxifraga rivularis s.l.*	10	5	1250	780	*
X	*Taraxacum arcticum*	840	1230	1230	780	670

ADT Altitude type
WA1-WA5 Study areas
* No specimens available
Altitudes in m

Greenland (Bay 1992). It grows in snow-patches, herb-slopes and sometimes in snow-covered heath (Böcher et al. 1978).

A. canescens was found at middle and higher altitude in WA1-WA3. It prefers sunny sites on gneissic or granitic rocks. It has been collected three times in WA1, north of Pingo Pass: 670-770 m. *A. canescens* is common in the Stauning Alper and has often been found at altitudes of at least 1000 m (WA2, east of Linné Gletscher: 1335 m, west of Linné Gletscher: 1030 m, 1120 m, Skjoldungebræ: 1310 m, 1320 m, Gullygletscher: 1230 m, Sefstrøm Gletscher:

1010 m, 1030 m, 1200 m, 1240 m, 1250 m, Schaffhauserdalen: 1080 m, Ismarken: 1000 m). In WA3 there are three collections above 600 m (WA3, Luciadal: 1060 m, Moræenedal: 700 m, 720 m).

Antennnaria porsildii ADT X (**11.3.2** N 4; **11.4.6; 11.4.8**)
A. porsildii is a tricentric, southern species with the northern limit at c. 74°20'N at the East coast (Bay 1992). This rare species has a similar distribution and similar ecological demands as *A. canescens.*

A. porsildii has been collected at two sites in WA1 and WA2 growing besides *A. canescens* (WA1, north of Pingo Pass: 670 m, WA2, Sefstrøm Gletscher: 1010 m). The specimens can easily be distinguished from *A. canescens*. This clear differentiation favours the hypothesis that the two taxa are apomictic small-species. This hypothesis is also sustained by the different chromosome numbers: 2 n = 56 for *A. canescens* and 2 n = 63 for *A. porsildii* (Böcher et al. 1978).

Arenaria pseudofrigida ADT X (**11.3.2** N 13; **11.4.10**)
A. pseudofrigida is a monocentric species found between Kap Dalton and Peary Land (Böcher et al. 1978, Bay 1992). The species grows on dry to humid sand or grus.

A. pseudofrigida was collected in WA1-WA4. In WA1, south of Randspids, *A. pseudofrigida* was found at seven sites in altitudes of 730 – 1100 m. There are more collections from WA1, Bersærkerbræ: 75 m, west of Kap Peterséns: 240 m. The species was also present in two other study areas (WA3, Agardh Bjerg: 1250 m; WA4, Waltershausengletscher: 450 m, Vibeke Sø: 350 m).

Armeria scabra ssp. *sibirica* ADT X (**11.3.2** N 9; **11.4.5;** Fig. 37)
A. scabra ssp. *sibirica* is found in Greenland south of c. 82°30'N, and is missing in some parts of Northwest Greenland (Bay 1992). It grows in the fell-field vegetation, in dry heaths, in shallow depressions kept wet during snowmelt but drying out in the summer (Böcher et al. 1978).

There is an interesting discussion about the ecological demands and the altitude distribution of this species. Gelting (1934) writes: "Occurs here and there throughout the area, sometimes in heaths, but mostly on bare gravel, especially near the beach. Often free of snow in the winter and hardly ever withstanding a prolonged snow-covering." He considers the species to be a halophyte and he suggests that observations of *A. scabra* ssp. *sibirica* at higher levels (up to 700 m) might be a relict on raised beaches from the time of the last big advance of the ice which lead to a rise in level of the sea. However, there might be an other explanation (Schwarzenbach 1960, 1961). In the most continental parts of the East Greenland fjords there are places with a crust of salts on the surface of the soil due to extreme evaporation of water from the uppermost layers above the permafrost. If such an accumulation of salt occurs in an area with gneissic or granitic bedrock the typical acidiphilous

Fig. 37. Armeria scabra ssp. sibirica. *Nathorst Land, niche N 9, Højedal (WA2, 900 m, 1954).*

species are not able to grow and are replaced by basiphilous species or by halophytes. So – strange enough – calcicolous species as *Dryas octopetala s.l.* or even a halo-phyte as *A. scabra* ssp. *sibirica* might be found at such sites because they can withstand a high content of mineral salts.

A. scabra ssp. *sibirica* has been collected in all five study areas WA1-WA5, but was rather rare. The following collections are from sites with a high content of mineral salts (WA1, Schuchert Dal: 100 m; WA2, Højedal: 900 m; WA3, Morænedal: 400 m; WA4, Krumme Langsø: 450 m).

Braya linearis ADT X (**11.3.2** N 10; **11.4.10**)
B. linearis has a bicentric distribution in Greenland with the northern limit at c. 74°N in East Greenland (Bay 1992). Böcher et al. (1978) write that *B. linearis* is most often found on dry soil rich in mineral salts, in grassy meadows and on "saltsøbredder". The species has similar ecological demands to *A. scabra* ssp. *sibirica*.

B. linearis has been collected four times in the western continental belt (WA3, east of Junctiondal: 1170 m, Moræenedal: 380 m, 400 m; WA4, Krumme Langsø: 20 m).

Carex bigelowii s.l. ADT X (**11.3.2** N 2, N 5, N 13; **11.4.8**)
Böcher et al. (1978) write that the northern limit of *C. bigelowii s.l.* is near 78°N. Therefore, it is doubtful, if the specimen from Kronprins Christian Land near Fyn Sø (80°30'N, 24°10'W, 20 m) determined by Kjeld Holmen belongs to this species. Bay (1992) has not included the record on his map showing the distribution of *C. bigelowii s.l.* and the specimen has been discarded.

C. bigelowii s.l. has been found in WA1-WA4. The species is common in the southern Werner Bjerge, in the Stauning Alper and in Nathorst Land. In WA3 and WA4 only a few specimens have been collected. The species has a similar altitude distribution as *Betula nana*.

Carex glacialis ADT X (**11.3.2** N 8, N 12, N 17; **11.4.10**)
C. glacialis has a tricentric distribution in Greenland (Bay 1992). At the eastern coast the species is found between Angmagssalik and 76°21'N as well as at Danmark Fjord, 81°10'N (Böcher et al. 1978). There are two more records from North Greenland (Fredskild et al. 1986). It was found also at Fligely Fjord, 74°50'N, 20°50'W (Fredskild & Bay 1990). The species grows on dry heath, and on dry places exposed to wind.

C. glacialis has been collected at five sites of middle and high altitude (WA2, Linné Gletscher: 1030 m, 1300 m; WA3, Benjamin Dal: 560 m, 650 m; WA5, Drømmebjerg: 650 m). This peculiar distribution of *C. glacialis* can not be explained on the base of the present knowledge .

Carex norvegica ADT X (**11.3.2** N 8, N 12; **11.4.10**)
C. norvegica has a southern distribution (Bay 1992) reaching 75°18'N at the eastern coast in Femdalen, Ostenfeld Land (Fredskild and Bay 1990). It grows in heaths, fens and herb-slopes.

C. norvegica has been found in three small areas only (WA1, Syltoppene: 210 m; WA2, Linné Gletscher: 920 m, 1030 m, 1120 m; WA3, Benjamin Dal: 650 m). The distribution of *C. norvegica* in WA2 and WA3 is similar to the pattern of *C. glacialis* .

Cochlearia groenlandica ADT X (**11.3.2** N 2; **11.4.6**; **11.4.7**)
C. groenlandica is a circumgreenlandic species (Bay 1992). Böcher et al. (1978) mention that the species grows at the sea-shore and around bird-perches where the soil is manured by the excrements of the birds. The same authors state that plants of *C. groenlandica* found in the inland could belong to a separate variety.

This species has been collected in WA1, Pingo Pass: 510 m and in the mountains N of Pingo Pass: 1000 m. These two sites are often frequented by geese, waders or other birds. It is therefore suggested that the propagules have been carried recently by birds to these sites. However, the population of *C. groenlandica* at 1000 m was very well established, so that the population might have survived at this high altitude for a long time. There are five specimens from Kronprins Christian Land (WA5, Ingolf Fjord: 5 m, Hekla Sund: 15 m, Rivieradal: 70 m, 70 m, 250 m).

Draba adamsii s.l. ADT X (**11.3.2** N 10)
D. adamsii s.l. has a northern distribution with an isolated occurrence in West Greenland (Bay 1992). It is found in East Greenland north of Scoresby Sund, growing on more or less humid soils and in snow-patches (Böcher et al. 1978).

D. adamsii s.l. has been collected in the sedimentary parts of WA1, WA3-WA5. It was not found in the Stauning Alper and in Nathorst Land. The species exceeds 1100 m in WA3 (Månesletten: 1285 m, Junctiondal: 1150 m, 1170 m). *D. adamsii s.l.* has been found three times in the nunataks north of 74° (WA4, nunatak of North Strindberg Land: 960 m, Waltershausengletscher: 450 m, C. H. Ostenfeld Nunatak: 720 m). Three specimens are from Kronprins Christian Land (WA5, Drømmebjerg: 600 m, Rivieradal: 70 m, 80 m).

Draba alpina s.l. ADT X (**11.3.2** N 1, N 10; **11.4.6**; **11.4.7**)
D. alpina s.l. has a tricentric distribution (Bay 1992). It occurs in East Greenland between 69°N and 77°N. The species grows on more or less humid ground, on solifluction soils, in snow-patches and on heaths (Böcher et al. 1978).

D. alpina s.l. has been found in WA1-WA5, 15 of the 22 specimens were collected in the southern Werner Bjerge around Pingo Pass. The species grows at low and middle altitude except in a few high sites in the sedimentary belt from the southern Werner Bjerge to East Andrée Land (WA1, north of Pingo Pass: 1000 m, 1200 m; WA3 Månesletten: 1180 m). One specimen has been collected in Kronprins Christian Land (WA5, Drømmebjerg: 600 m).

Draba arctogena ADT X (**11.4.10**)
Bay (1992) has revised the material of *D. arctogena* kept at the Greenland Herbarium Copenhagen. As his map shows, *D. arctogena* has a northern distribution and an isolated occurrence in West Greenland. It reaches the southern limit in East Greenland at 74°N. The species grows in the fell-field vegetation and in snow-patches (Böcher et al. 1978).

D. arctogena has been collected in WA3 and WA4 at altitudes of 900 m or more (WA3, Gauligletscher: 1150 m; WA4, western Bartholin Land: 1370 m, North Strindberg Land: 960 m, Vibeke Nunatak: 900 m).

Draba gredinii ADT X (**11.3.2** N 1; **11.4.7**)
D. gredinii is found in Northeast Greenland. The species grows on solifluction soils (Böcher et al. 1978).

There are only two specimens determined as *D. gredinii* (WA1, north of Pingo Pass: 1200 m; WA4, North Strindberg Land: 960 m).

Empetrum nigrum ssp. *hermaphroditum* ADT X (**11.3.2** N 4; **11.4.8**)
E. nigrum ssp. *hermaphroditum* is a southern species (Bay 1992) with the northern limit in East Greenland at 76°46'N (Böcher et al. 1978). The species is a typical plant of dry heath at low altitude, and is often found together with *Betula nana*.

E. nigrum ssp. *hermaphroditum* was collected in WA1, Syltoppene: 20 m, Flødegletscher: 210 m, 250 m, Bersærkerbræ: 340 m. The species has been found in WA2 at low levels with a few exceptions (WA2, Sefstrøm Gletscher: 380 m, north of Vikingebræ: 840 m, 870 m, Tærskeldal: 350 m). There are a few sites at considerable altitude in WA3, Benjamin Dal: 440 m, 550 m, Luciadal: 470 m.

Erigeron compositus ADT X (**11.3.2** N 8, N 12; **11.4.10**)
E. compositus has nearly a circumgreenlandic distribution, missing only in Northwest Greenland (Bay 1992). There are two chromosomal races with 2n = 54 and 2n = 63. The species grows on dry sunny slopes and in rocks (Böcher et al. 1978). *E. compositus* was collected only in two study areas (WA2, Linné Gletscher: 1150 m; WA3, Benjamin Dal: 1150 m).

Festuca hyperborea ADT X (**11.3.2** N 1, N 10, N 17)
The northern species *F. hyperborea* has the southern limit near Scoresby Sund (Bay 1992. The species grows on open humid soils with sand or clay, on slopes, on solifluction-soils and in snow-patches (Böcher et al. 1978).

F. hyperborea was found in WA1, WA3-WA5, following the belt of sediments. It is a species of middle and high altitude. Five specimens have been collected at altitudes above 900 m (WA1, north of Pingo Pass: 1200 m; WA3, east of Junctiondal: 950 m, 1140 m, 1150 m, Gauligletscher: 1150 m). There is only one specimen from WA4, Vibeke Sø: 230 m. *F. hyperborea* is common in WA5 reaching rather high altitudes (WA5, Drømmebjerg: 670 m, Ingolf Fjord: 470 m).

Festuca vivipara ADT X (**11.3.2** N 10; **11.4.6**)
F. vivipara is a monocentric species found in Greenland between Scoresby Sund (70°N) and Peary Land, 83°N (Bay 1992). It has several races with different numbers of chromosomes (Böcher et al. 1978).

This species was found in the study areas WA1-WA4. The majority of specimens have been collected in altitudes between 350 m and 770 m. There are three exceptions (WA1, north of Pingo Pass: 1100 m; WA3, east of Junctiondal: 1150 m, upper Junctiondal: 1100 m).

Huperzia selago ADT X (**11.3.2** N 4; **11.4.8; 11.4.10**)
H. selago is nearly a circumgreenlandic species, missing only in North Greenland (Bay 1992). The species has been found as far north as Campanuladal, 80°38'N (Bay & Fredskild 1994). It grows in heaths and on rocks at places where there is late snow (Böcher et al. 1978).

H. selago has been collected in WA1-WA3. It was often found together with *Cassiope tetragona* at altitudes not exceeding 510 m with the exception of three high level collections (WA2, Sefstrøm Gletscher: 1030 m; east of Skjoldungebræ: 1240 m, Tærskeldal: 690 m).

Luzula spicata ADT X (**11.3.2** N 4, N 7, N 8, N 12; **11.4.6**; **11.4.8**)
The southern species *L. spicata* reaches c. 76°30'N at the East coast of Greenland. It grows in heaths, in dry to middle-humid grassy meadows, in herb-slopes and lichens (Böcher et al. 1978).

L. spicata has been collected in WA1-WA3, but not in WA4. The species is typical of the alpine herb-slopes. *Luzula spicata* often grows together with *Antennaria canescens* and *Arnica angustifolia*. It has not been found on calcareous or dolomitic sediments.

There is a more or less closed area at high altitude in the Stauning Alper (WA2, Skjoldungebræ: 1450 m, 1660 m, eastern side of Linné Gletscher: 900 m, 1335 m, western side of Linné Gletscher: 1020 m, 1100 m, Sefstrøm Gletscher: 960 m, 1010 m 1200 m, 1260 m, north of Vikingebræ: 810 m). It seems possible that this population belongs to a well adapted variety of *L. spicata* (11.4.6). This hypothesis is sustained by the fact that Böcher et al. (1978) mention populations with different chromosome numbers: 2n = 24 (12, 14, 18). *L. spicata* has also been collected in WA3, Benjamin Dal: 650 m.

Minuartia rossii ADT X (**11.3.2** N 12, N 17)
The northern species *M. rossii* has its southern limit at the East coast at Ymers Ø, 73°10'N (Böcher et al. 1978). The species is found on wet sand and grus.

M. rossii was collected in WA3 (Benjamin Dal: 650 m, Morænedal: 380 m). It was quite common in Kronprins Christian Land with the following maxima (WA5, Drømmebjerg: 770 m, Ingolf Fjord: 150 m).

Phippsia algida s.l. ADT X (**11.3.2**, N 10, N 13, N 17; **11.4.6**)
Bay (1993) discusses the taxonomy and the geographical distribution of *P. algida s.l.* and of *P. algida* ssp. *algidiformis*. The species is found in East and Northeast Greenland north of Scoresby Sund (Bay 1992). *P. algida s.l.* is a typical plant of late snow-patches, of constantly wet fields and of solifluction soils.

This species was collected in all five study areas. The altitude distribution shows a peculiar pattern which might be due to a segregation of ecologically different populations.

It is equally distributed in WA1, Pingo Pass between 300 m and 620 m. Further, *P. algida s.l.* occurs in a small valley of the Syltoppene: 620 m. There are many sites at high altitude in the sedimentary parts of WA2 – WA5 (WA2, Højedal: 800 m; WA3, Månesletten: 1285 m, east of Junctiondal: 1150 m, 1160 m, Gauligletscher; 1150 m, Moraenedal: 1180 m, Agardh Bjerg: 1250 m; WA4, Bartholin Land: 1180 m, 1200 m; WA5, Drømmebjerg: 700 m). It would be interesting to study whether or not the specimens of these regions belong to segregated populations of *P. algida s.l.*

Poa alpina s.l. ADT X (**11.3.2** N 3, N 10; **11.4.6**)
P. alpina s.l. is a southern species reaching the northern limit at c. 75° (Bay 1992). The species grows in heaths, on herb-slopes and, by early melting snow-patches. The viviparous variety (*P. alpina* var. *vivipara*) is found in East Greenland between 68°31'N and Clavering Ø, and is recorded also from Ingolf Fjord (80°35'N). It grows on solifluction soils (Böcher et al. 1978). Böcher et al. (1978) mention different chromosome numbers with 2n = 28, 33, 42-45, 46 (14-74) for the non-viviparous *Poa alpina*. It is probable that there are populations segregated by genetic barriers.

Schwarzenbach (1956) has collected *P. alpina* var. *vivipara* near the settlement of Scoresby Sund. He showed by a cultivation experiment that the viviparity depends on the influence of frost and photoperiodism.

Schwarzenbach (1961) distinguishes a variety *P. alpina* var. *saxicola* from the common *P. alpina* var. *typica*. He assumes that the variety *saxicola* would only be found at high levels between 72°N and 74°N.

P. alpina var. *typica* has been collected in WA1-WA4. It was found in WA1 at low and middle altitude (WA1, Biskop Alf Dal: 840 m, Syltoppene: 570 m). The species grows in WA2 at low altitude along the Segelsällskapet Fjord and the Alpefjord. *Poa alpina* var. *typica* reached 700 m in WA3, Moraenedal. It was collected in WA4, Krumme Langsø at altitudes up to 260 m. The variety *P. alpina* var. *saxicola* was collected at several sites (WA2, east of Skjoldungebræ: 1330 m, north of Vikingebræ: 810 m, Sedgwick Gletscher: 780 m; WA3, Gauligletscher: 1140 m; WA4, Vibeke Nunatak: 920 m).

Poa pratensis s.l. ADT X (**11.3.2** N 10; **11.4.6**)
P. pratensis s.l. (excluding the northern viviparous *P. pratensis* var. *colpodea*) is a southern species reaching c. 75°N on the East coast (Bay 1992). The species is found in heaths and on herb-slopes.

P. pratensis s.l. (including *var. colpodea*) has been collected in all five study areas. It is a species of low and middle altitude, found below 450 m with a few exceptions (WA1, Pingo Pass: 520 m; WA2, Schaffhauserdalen: 710 m; WA3, Månesletten: 1185 m, Moraenedal: 700 m; WA4: Waltershausen Nunatak: 450 m).

Potentilla hyparctica ADT X (**11.3.2** N 13, N 14, N 16, N 17; **11.4.6**)
The northern species *P. hyparctica* is found on solifluction soils, on grus and sometimes around bird-perches (Böcher et al. 1978). *P. hyparctica* often flowers early.

P. hyparctica has been collected in WA2-WA5. There are two sites of remarkable altitude (WA2, north of Vikingebræ: 810 m; WA3, Agardh Bjerg; 1350 m). Seven collections are from WA4, where the nunataks north of 74°N could be a niche for a small-species well adapted to the continental climate near the Inland Ice (see 11.4.6). The species was found in WA5 in two sites with the same geographical position (WA5, Drømmebjerg: 80°18'N, 21°26'W, 670 m.

Pyrola grandiflora ADT X (**11.3.2** N 4, N 13; **11.4.10**)
P. grandiflora is a southern species with the northern limit at 74°30'N in East Greenland (Böcher et al. 1978). This species is found on dry heaths and in scrub, sometimes also on dry slopes, in screes and around bird-perches.
This species has been collected in WA1-WA4 at low and middle altitude. Most sites are below 500 m with a few exceptions (WA1, Pingo Pass: 540 m; WA2, Sefstrøm Gletscher: 960 m, north of Vikingebræ: 870 m; WA3, Morænedal: 970 m, Agardh Bjerg: 850 m; WA4, Vibeke Sø: 540 m). *P. grandiflora* shows a trend to establish populations with a large number of individuals on numerous sites within small areas.

Ranunculus glacialis ADT X (**11.3.2** N 13, N 14; **11.4.10**, Fig. 38)
R. glacialis is a monocentric species in East and Northeast Greenland (Bay 1992). It is found northwards to c. 78°33'N (Fredskild & Bay 1990). The species grows by late snow-patches and on constantly soaked grus (Fig.38).

R. glacialis was found in WA3 and WA4. The seven collections mark an area in the westernmost belt between 73°46'N and 74°27'N in altitudes from 780 m and 1400 m (WA3, Agardh Bjerg: 1100 m; WA4, nunatak in North Strindberg Land: 960 m, Waltershausen Nunatak: 1270 m, Bartholin Land: 900 m, 1350 m, Vibeke Nunatak: 900 m, Vibeke Sø: 780 m).

Ranunculus pygmaeus ADT X (**11.3.2** N 4; **11.4.10**)
R. pygmaeus is found northwards to c. 79°N (Bay 1992). The species is restricted to snow-patches and solifluction-soils.

R. pygmaeus was collected in WA1-WA3 generally at altitudes below 600 m but with some exceptions (WA1, Pingo Pass: 620 m, Syltoppene: 620 m; WA2, Sefstrøm Gletscher: 1010 m, 1200 m; WA3, Moræenedal: 650 m). *R. pygmaeus* was not found on the nunataks of WA4.

Sagina intermedia ADT X
The circumgreenlandic species *S. intermedia* (Bay 1992) grows on sand and clay, sometimes also by snow-patches (Böcher et al. 1978). Five specimens

Fig. 38. Ranunculus glacialis. *Vibeke Sø (WA4, 780 m, 1952).*

have been collected (WA1, Biskop Alf Dal: 690 m; WA2, Forsblad Fjord, Caledonia Ø: 5 m, south of Gully Gletscher/Sefstrøm Gletscher: 10 m; WA4, Vibeke Sø: 780 m; WA5, Rivieradal: 70 m).

Salix herbacea ADT X (**11.3.2** N 4; **11.4.10**)
S. herbacea has a southern distribution reaching 76°46'N in East Greenland (Böcher et al. 1978). The species is found by late-melting snow-patches and on herb-slopes of low altitude.

S. herbacea has been collected three times (WA2, Sefstrøm Gletscher: 1030 m, northern side of Forsblad Fjord: 20 m; WA3, Moraenedal: 420 m).

Saxifraga hieracifolia ADT X

The conspicuous species *S. hieracifolia* has a bicentric distribution (Bay 1992). It is found in East Greenland from 70°N to 73°35'N and in Peary Land (Böcher et al. 1978). It grows by snow-patches and in mosses. *S. hieracifolia* was collected in both sub-areas of WA1 (Pingo Pass: 520 m; Syltoppene: 110 m, west of Kap Peterséns: 70 m).

Saxifraga rivularis s.l. ADT X (**11.3.2** N 10)
S. rivularis s.l. has a southern distribution (Bay 1992) and has been found at the East coast to c. 78°N (Böcher et al. 1978). The species grows near running water, in constantly irrigated polygon-fields and in mosses.

There are only six sites in WA1-WA4, where *S. rivularis s.l* was collected (WA1, Kap Petersén̈s: 10 m; WA2, Forsblad Fjord, Caledonia Ø: 5 m, Forsblad Fjord, head of the fjord: 1 m; WA3, east of Junctiondal: 1250 m, Moræ̈nedal: 480 m; WA4, Vibeke Sø: 780 m). .

Taraxacum arcticum ADT X (**11.3.2** N 4, N 10, N 17; **11.4.6**)
T. arcticum is found from East to North Greenland (Bay 1992). It grows in snow-patches, on wet ground, on herb-slopes and sometimes in the fell-field vegetation (Böcher et al. 1978).

The species was found in all five study areas. It was rather common in WA1, Pingo Pass with the highest site at 840 m. Four collections are from the other study areas (WA2, Sefstrøm Gletscher: 1230 m; WA3, east of Junctiondal: 1230 m; WA4, Vibeke Sø: 780 m; WA5, Drømmebjerg: 670 m).

7. Comparison of ADT with North Greenland Distribution types (NGDT)

7.1 Introduction

Several authors have discussed the hypothesis that there might be some relationship between the geographical and altitude distribution of species in Greenland (Gelting 1934, Böcher 1963, Gribbon 1968, Halliday (pers. comm.)

Gelting (1934) finds some relationship between the vegetation of the outer coast and the vegetation from the zone of persistent snow-patches ("Nival-zone"). Gribbon (1968) has found that in Southeast Greenland the percentage of low-arctic oceanic montane species decreased with increasing altitude. This was paralleled by the decrease in the smaller group of widespread low- and middle-arctic species. Corresponding with these declines is the increasing percentage of widespread arctic montane species which dominate the flora at higher altitudes. Gribbon (1968) also noted that the two small groups of widespread arctic continental species and medium-arctic montane species were appreciably more common between 1000 m and 1600 m, suggesting a relationship between an increase in altitude and an increase in latitude or continentality. Halliday (pers. comm.) has analysed Stocken's data in the Report of the Royal Navy East Greenland Expedition 1966, and his results essentially confirm Gribbon's conclusions.

The collections from the study areas offer the opportunity to study the relationships between the geographical distribution of species in Greenland and their altitude distribution. As basis for comparison within the study areas the 14 "North Greenland distribution types (NGDT)" proposed by Bay (1992: 19) have been chosen. Though he considers only the area north of 74°N his types of geographical distribution can also be used to describe the species distribution between 71°N and 74°N. The 14 NGDT types provide for a fairly detailed classification of the 176 species.

7.2 Representation of the types NGDT in the study areas

The collected species in each study area (WA1-WA5) have been grouped according to the NGDT-types defined by Bay (1992). His 14 main NGDT types are shown in Fig. 39. It should be noted that they take no account of altitude. Bay's subtypes have also been considered, but due to the small numbers of species in some of these this subclassification has been omitted. The results are shown in Tab. 32.

Eleven of the fourteen main NGDT types are present in at least one of the

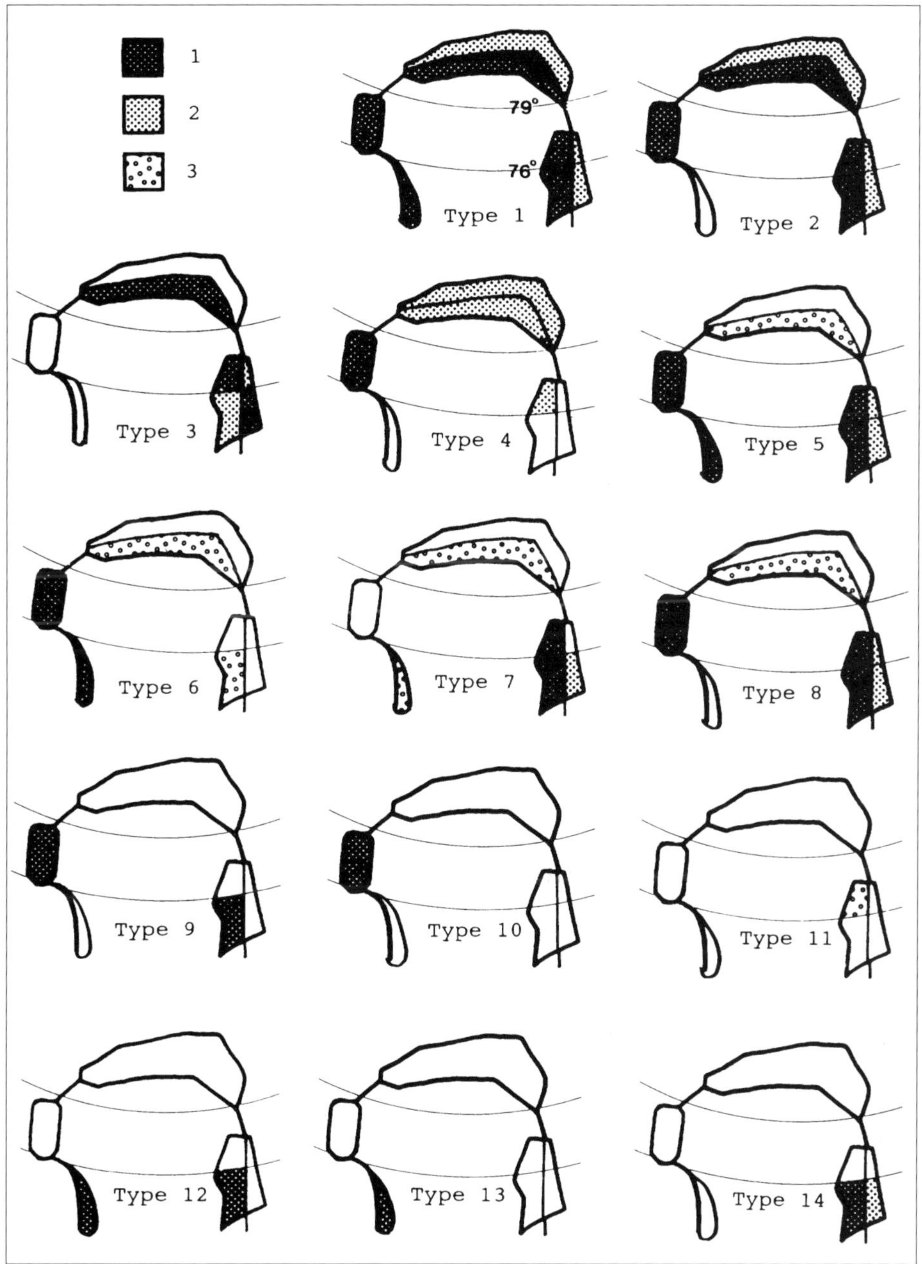

Fig. 39. Distribution types in northern Greenland. The hatched and dotted areas indicate that the main type includes subtypes. 1: continuous distribution, 2: a few species missing, 3: isolated occurences of a few species. (The original figure has been published by Bay 1992).

five study areas. Types 1-3 and 5-8 are found in all areas WA1-WA5. Type 4 (species only found north of 76°N) has been observed in WA5 only. Type 9 is represented by the species *Arenaria humifusa* in WA1 only. The species of the two types 12 and 14 (found only south of 76°N) are missing in WA5. Types 10, 11 and 12 are missing in all study areas.

Tab. 32. Percentage of 'Northern Greenland distribution types NGDT (Bay 1992)' within the study areas WA1-WA5 (herbarium specimens only).

NGDT	WA1 %	WA2 %	WA3 %	WA4 %	WA5 %	WA1-5 %
1	31.0	34.2	32.0	36.6	48.4	27.3
2	12.4	7.6	15.6	16.3	23.7	13.6
3	0.8	0.8	0.8	0.8	4.3	2.3
4	–.-	–.-	–.-	–.-	2.2	1.1
5	13.2	18.3	16.4	13.0	7.5	13.1
6	2.3	3.3	2.3	2.4	2.1	2.8
7	7.7	5.0	7.8	6.6	2.1	7.4
8	10.9	10.8	12.6	11.4	9.7	10.2
9	0.8	–.-	–.-	–.-	–.-	0.6
12	4.6	5.0	3.9	3.2	–.-	3.4
14	13.2	10.8	7.8	8.1	–.-	12.5
xxx	3.1	4.2	0.8	1.6	–.-	5.7
sum	100.0	100.0	100.0	100.0	100.0	100.0
species	129	120	128	123	93	176

Abbreviation:
xxx NGDT uncertain

The percentages of the three types NGDT 1-3 increase from WA1 to WA5 with two exceptions (NGDT 1 in WA3, NGDT 2 in WA2). This general trend from south to north might be shown more clearly by aggregating the NGDT types 1-4 to a new group "species found in the northern part of Greenland". This is set out in Tab. 33.

There is a decrease from left to right of the aggregate of the types 12 and 14. The following Tab. 34 shows the trend if these two types are aggregated into a new group "species found in the eastern part of Greenland".

7.3 Altitude limits of the types of distribution (limits based on all 9626 observations in WA1-WA5)

A second method has been used to study the altitude distribution of Northern

Tab. 33. Distribution of the aggregated types NGDT 1-4

WA1	WA2	WA3	WA4	WA5
44.2%	42.6%	48.4%	53.7%	78.6%

Tab. 34. Distribution of the aggregated types NGDT 12 + 14

WA1	WA2	WA3	WA4	WA5
17.8%	15.8%	11.7%	11.3%	0.0%

Tab. 35. Average altitudes of frequent 'Northern geographical distribution types (NGDT)' within the study areas (9626 observations)

NGDT	WA1 m hgt	WA2 m hgt	WA3 m hgt	WA4 m hgt	WA5 m hgt	WA1-5 m hgt
1	624	743	661	790	293	652
2	599	700	716	727	222	574
5	465	473	568	640	145	504
6	344	306	670	327	201	379
7	552	669	592	546	157	576
8	463	454	486	500	143	449
12	357	641	429	403	–.-	475
14	435	321	353	389	–.-	381
Mean	479	538	559	540	145	424

WA1-5 Average of altitudes for the individual NGDT in the total area of WA1-WA5
Mean Average of altitudes within the columns WA1-WA5 and WA1-5

Greenland distribution types (NGDT) in WA1-WA5. The analysis has been done on the 3457 herbarium specimens, as well as on all 9626 observations. The database "9626 observations" is chosen as an example, as the larger material gives a more accurate estimation of the altitude averages of the different types.

The average altitude limits of all species belonging to different NGDT in the five study areas and in the total area (WA1-WA5) are shown in Tab. 35. Within the columns there is a trend of decreasing altitude from top to bottom, but with different levels for each column. There is an irregularity in WA3. The average of NGDT 2 (716 m) is higher than the average of NGDT 1 (661 m), due to some high records of some species of NGDT 2 in WA3.

7.4 Altitude distribution of the three aggregated groups of distribution based on NGDT in WA1-WA5

a) Type "northern distribution (N)" aggregated of NGDT 1, 2, 3 and 4.
b) Type "north-eastern distribution (NE)" aggregated of NGDT 5, 6, 7 and 8.
c) Type "eastern distribution (E)" aggregated of NGDT 12 and 14.

Types with northern distribution "N" (NGDT = 1,2,3 and 4) as found in WA1-WA5

There are peaks in numbers of species at medium altitude within the columns of WA1 (450 m), WA3 (750 m), and WA4 (1050 m) as shown in Tab. 36 a. WA5 has a peak in the lowest band (150 m). WA2 has two peaks at 150 m and 1050 m: The first peak at 150 m is due to the exceptionally high number of records along the coast of Forsblad Fjord.

The highest means are found in the two continental study areas WA2 (741

Tab. 36. Altitude distribution of the most frequent 'Northern Greenland distribution types (NGDT)' within the study areas WA1-WA5

a) Types with northern distribution 'N' (NGDT = 1,2,3 and 4) as found in WA1-WA5

alt	WA1	WA2	WA3	WA4	WA5	WA1-5
m	n	n	n	n	n	n
1650	–.-	26	3	–.-	–.-	29
1350	23	77	33	62	–.-	195
1050	59	91	103	154	1	408
750	157	44	145	130	91	567
450	167	8	135	69	41	420
150	68	132	71	64	211	546
n	474	378	490	479	344	2165
mean	619	741	675	775	268	632

Explanations

alt	altitude band with a range of 300 m
m	meters
WA1-WA5	study areas
WA1-5	total sum of all five study areas
n	number of records based on 9626 observations
mean	average of altitudes

b) Types with north-eastern distribution 'NE' (NGDT = 5,6,7 and 8)

alt	WA1	WA2	WA3	WA4	WA5	WA1-5
m	n	n	n	n	n	n
1650	–.-	5	–.-	–.-	–.-	5
1350	3	28	11	5	–.-	47
1050	15	49	23	33	–.-	120
750	91	30	58	21	6	206
450	104	12	88	37	2	243
150	86	137	41	49	35	348
n	299	261	221	145	43	969
mean	480	489	550	556	151	495

c) Types with eastern distribution 'E' (NGDT = 12 and 14)

alt	WA1	WA2	WA3	WA4	WA5	WA1-5
m	n	n	n	n	n	n
1650	–.-	–.-	–.-	–.-	–.-	–.-
1350	3	28	11	5	–.-	47
1050	15	49	23	33	–.-	120
750	91	30	58	21	–.-	200
450	104	12	88	37	–.-	241
150	86	137	41	49	–.-	313
n	299	256	221	145	–.-	921
mean	404	463	386	394	–.-	418

m) and WA4 (775 m), and the lowest in WA5 (268 m). The mean in the column T (WA1-WA5) is 632 m. It seems that the group of species with northern distribution in the study areas WA1-WA4 prefers high and middle altitudes.

A similar pattern is found for the group NE (Tab. 36 b). The numbers of records are lower in the bands of 1350 m and 1650 m than in Tab. 36 a. This is an indication, that the species of the group NE find their optimum at lower levels compared with the species of the group ‚N'.

The means of WA1-WA4 have a small range of only 76 meters. As expected, the mean of WA5 (151 m) is very low. The mean of the column T (WA1-WA5) has a value of 495 m, 137 m lower than in the group N. This difference suggests that the group ‚NE' requires a higher temperature (appr. 1°C) during the growing season corresponding to a longer period of temperature above zero (1-2 weeks).

Species of the group ‚E' (Tab. 36 c) are not found in WA5, which is to be expected in view of the criteria for NGDT 12 and NGDT 14 (Bay 1992). Within the four study areas WA1-WA4 there are no records from the altitude band 1650 m, and only 47 from altitude band 1350 m.

The means of WA1-WA4 have a small range with values of 386 m to 463 m. The mean of column T (WA1-WA5) has a value of 418 m, 214 m below the mean of the group ‚N' and 77 m below the mean of the group ‚NE'. The difference between the groups ‚NE' and ‚E' is rather small and might be explained by the fact that some species of the group ‚E' have been found at surprisingly high levels in WA1-WA4.

In a next step the averages of the four means WA1-WA4 for the three groups have been calculated. (‚N': 703 m, ‚NE': 519 m and ‚E': 404 m). The differences between the three groups are now more evident: (‚N'/‚NE': 184 m, ‚NE'/‚E': 115 m, ‚NE'/‚E': 299 m).

7.5 Distribution of the types NGDT within the highest 300 m band of all five study areas

The results of the analysis of all specimens collected from the highest altitude band of the five study areas are shown in Tab. 37. This highest band has a range of 300 m. The upper limit is given by the altitude of the highest part of the area where plants have been collected.

All records from the highest 300 m of WA1-WA5 have been stored in a new database with a total of 366 specimens comprising 80 species. Tab. 37 shows the frequencies of the species from the main NGDT. Most frequent are species of NGDT 1 (48.7%) and NGDT 2 (20.0%). If the four geographical types NGDT 1-4 are aggregated to a new group ‚N', the percentage reaches 72.4%. The values for the four types NGDT 5-8 (group ‚NE') are much lower giving a total sum of 26.4%. The group ‚E' is not represented in the highest altitude

Tab. 37. Distribution of the 'Northern Greenland distribution types (NGDT)' within the highest 300 m of all five study areas

Group	NGDT	observations	%
N	1	39	48.7
N	2	16	20.0
N	3	2	2.5
N	4	1	1.2
NE	5	7	8.8
NE	6	2	2.5
NE	7	7	8.8
E	8	5	6.3
E	12	–.-	–.-
E	14	–.-	–.-
	xxx	1	1.2
	Sum	80	100.0

Abbreviations
N Northern
NE North Eastern
E Eastern

zone of the five study areas. In conclusion this shows that the highest altitude zone has only species of the two groups ,N' and ,NE' split by a ratio of 3:1 respectivily.

7.6 Comparison between ADT and NGDT

The introduction of this classification system of altitude distribution types makes it possible to compare the ADT with the NGDT defined by Bay (1992). The altitude distribution types of species compared with Bay's system of "Northern Greenland Distribution Types (NGDT)", and for the 132 species classified by the ADT-system are shown in Tab. 38.

The NGDT 1-4 are present in the altitude classes 1 and 2 with three exceptions in ADT C3 and ADT D3. On the other hand the NGDT 12 and 14 are only present in the altitude classes 2 and 3 (B2, D2, B3, C3, G3). This result supports the hypothesis that species with a northern distribution in Greenland reach higher altitudes in the mountainous parts of Northeast and East Greenland than species with latitude limits further to the south.

7.7 Conclusion

The hypothesis of several authors (Gelting 1934, Böcher 1963, Gribbon 1968, Halliday (pers. comm.) is confirmed that there is a relationship between the

Tab. 38. Comparisons between ADT and NGDT (only main-types) in all study areas WA1-WA5

ADT	NGDT												
	1	2	3	4	5	6	7	8	9	12	14	S1	S2
A1	5	2	–	–	1	–	–	–	–	–		8	
B1	15	3	–	–	2	–	–	1	–	–	–	21	
C1	4	1	–	–	–	–	–	–	–	–	–	5	
D1	1	–	–	–	–	–	–	–	–	–	–	1	
H1	–	1	1	1	–	–	1	–	–	–	–	4	39
A2	4	1	–	–	1	–	3	2	–	–	–	11	
B2	5	2	–	–	5	–	–	–	–	–	1	13	
C2	3	2	–	–	1	–	–	1	–	–	–	7	
D2	1	3	–	–	–	–	–	3	–	–	1	8	
H2	2	3	2	1	–	–	1	–	–	–	–	9	48
B3	–	–	–	–	1	–	1	1	–	4	4	11	
C3	1	2	–	–	1	–	1	–	–	–	1	6	
D3	2	–	–	–	1	–	–	2	–	–	–	5	
E3	–	–	–	–	–	–	1	2	–	–	–	3	
F3	–	–	–	–	2	1	–	1	–	–	–	4	
G3	–	–	–	–	1	–	–	1	1	1	12	16	45
S1	43	20	3	2	16	1	8	14	1	5	19	132	132

Abbreviations

NGDT	Northern Greenland distribution type (Bay 1992)
ADT	Altitude distribution type
S1	Sum of species on NGDT geographical types
S2	sum of altitude classes altitude classes and habitat classification (ADT)
A-H	Habitat classes
1-3	Altitude classes (ADT)

geographical and the altitude distribution of vascular plants in East and Northeast Greenland. It is shown that the group N (NGDT 1-4) reaches the highest levels in all study areas. The species of the group NE (NGDT 5-8) are found at somewhat lower altitudes. The species of the group E (NGDT 12 and 14) grow at the lowest levels and have not been found in the northernmost study area WA5. The differences between the altitude distributions of the three main groups (N, NE, and E) are explained either by the climatic conditions, reflecting both decrease of mean temperature during summer with increasing altitude, or by increase of temperature in continental parts of East and Northeast Greenland.

8. Estimation of the altitude limit of vascular plants

8.1 Definition

The expression "altitude limit of vascular plants" is defined as the altitude of the highest observation of a vascular plant within a geographically delineated area. It is synonymous to the German word "Vegetationsgrenze" often used in publications on the vegetation of the Alps.

8.2 Ecological interpretation of the altitude limit of vascular plants

The temperature of the air and of the soil decreases with increasing altitude. As the temperature during the growth-season is a limiting ecological factor a species can reach only a certain altitude. Therefore, the specific altitude limits of vascular plants might be used to estimate the minimum level of the local temperature conditions during the growing season. If the sequence of the altitude limits of the species is known, the relative position allows an estimation of the minimal temperature demanded by the individual species. If it is possible to relate the altitude limits by defined isotherms, a map of the altitude limit of vascular plants offers a chance to compare the conditions of the summer-temperature in different regions.

Two altitude limits are often used as indications for temperature conditions. The "tree-line"shows the highest sites of trees observed in the field not distinguishing between species. The "altitude limit of vascular plants" links the highest sites of vascular plants not distinguishing between species. However, a few points must be kept in mind using the altitude limits of plants as a bioindication of temperature. Botanists studying the ecology of alpine and arctic plants agree that the development and the growth of plants is controlled to a large extent by the local micro-conditions of climate and soil. These may change within very small distances as small as millimeters or centimetres. The existence of a species at a given site depends on many interacting factors. The following have been observed in the Stauning Alper:

- Drifting snow accumulates behind ridges. This snow-cover delays the development of plants. The growing season will be shortened considerably and species requiring a long period of days with positive temperatures can not persist in such conditions. However, some plants can start growing

very quickly, even after being covered with snow, for example *Saxifraga oppositifolia*.
- The hours of sunshine are reduced on steep north-facing slopes inhibiting, that the soil is warming.
- The small hummocks of the arctic tundra show strong gradients of soil temperature from top to bottom.
- The soil of a sunny crest exposed to local winds is rather cool in spite of the many hours of sunshine. The wind sweeps away the snow-cover as well as the fine material, and creates very dry conditions which inhibit plant growth.
- Moisture availability is crucial to many plants.

Nevertheless, the influence of the micro-conditions of the habitat is limited and the macro-climate still controls the level of the altitude limits.

8.3 Methods

8.3.1 Introduction

The levels of the "tree-line" and of the "altitude limit of vascular plants" are often determined by the highest observations within the studied area. The literature about the altitude distribution of plants shows that this way of determination has been accepted without evaluating the reliability of this method.

Therefore, a study comparing of several methods has been done in interdisciplinary co-operation with W. Urfer, R. Geißdörfer (1994) and S. Kaiser, Department of Statistics of the University of Dortmund (Germany). The analysis has been based on the botanical observations in the mountains of East and Northeast Greenland. The database (situation 1993) differs slightly from the data of the present study since it includes 9348 instead of 9626 observations and 180 instead of 176 species. The difference is explained by two facts: A small group of critical species has been revised and these changes are included after 1993. Further, a few small-species have been aggregated to broader taxonomic units, before the data for the present study was analysed. As mentioned in chapter 8.3.4 the statistical analysis of the database 1993/94 define nine geographical ‚provinces' instead of the five study areas WA1-WA5 (see Tab. 39).

The above statistical study (1993/94) concentrates on the estimation of the altitude limit of vascular plants. This approach is acceptable as the same methods could be used for the estimation of the altitude limit of any single species.

Two types of statistical methods are compared. The first type is based on direct observations in the field above and below the altitude limit of vascular plants. The second type uses all observations within a ‚province'. The decrease in numbers of observations, in sites and in species with increasing altitude is described as a mathematical function, which makes it possible, to estimate the altitude limit of vascular plants by extrapolation. It is assumed that the altitude limit is reached, when the number of observations (sites and species) converges towards zero. Preliminary results of this interdisciplinary approach have been published by Urfer & Schwarzenbach (1995).

8.3.2 Determination by type 1 method (highest observations in the field)

The simplest method is to record the highest altitude where a vascular plant has been observed, and using this figure to estimate the altitude limit of vascular plants within a geographically defined area.

This estimation needs interpretation. Whenever a vascular plant has been found at a certain site the altitude limit of vascular plants must be at least as high as the altitude of this site. There is however no proof that this recorded site is the highest location of the vascular plants.

When doing field work it is difficult to find the highest site where species of vascular plants might grow, for the following reasons:

- The theory of sampling indicates that even a small sample might include a remote record lying far away from all the other records. Outliers are rare, but they cannot be excluded.
- Ecological niches enabling plants to grow become rare near the altitude limit of vascular plants. The likelihood of finding such outposts is therefore low.

The following example further illustrates this last point. The five highest sites of vascular plants (based on all observations) in a region (WA2, Stauning Alper, innermost Skjoldungebræ) show the following altitudes: 2200 m, 2050 m, 1970 m, 1870 m, and 1720 m. The range (480 m) is exceptionally high due to the unexpected observation of a flowering *Saxifraga cernua* in a south-facing fissure of the summit-crest of Elisabeth Bjerg (2200 m).

The reliability therefore of the highest observation as an estimation of the altitude limit of vascular plants depends on the intensity of observation in the area which is 200-300 m above the highest site previously recorded. A careful search can lead to the result that the third, fourth and fifth altitude of the sequence of the five highest records have a range of less than 100 m. Therefore, these three altitudes can give a fairly reliable estimation of the altitude limit of vascular plants, provided the field work includes levels above the assumed altitude limit.

8.3.3 Determination by the regression "rate of decrease of the number of species/100 m altitude" (type 2a method)

Dr G. Halliday (pers. comm.) has studied the altitude distribution of vascular plants analysing the mainly unpublished data of many collectors for the area between Angmagssalik and the Blosseville Kyst (66°-69°N) in Southeast Greenland. He compares his results with observations of Gribbon (1968) and Schwarzenbach (1961) in other areas of Greenland.

Halliday suggests the "rate of decrease of the number of species/100 m altitude" as a simple parameter for comparison of the altitude distribution of vascular plants. These rates are calculated for the altitude sequence of the most species rich sites. Such sites are mostly found in places with a favourable insolation and moderate humidity. These rates of decrease range from 2.1 species/100 m rise (Angmagssalik) and 6.6/100 m rise (Kap Farvel). A rate of 3.3 species/100 m rise has been calculated by him for the nunataks north of 74°N (WA4).

Halliday has calculated the function ‚decrease of number of species with increasing altitude' based on a linear regression model. Using this function it is possible to estimate the altitude limit of vascular plants by extrapolation. This estimation for a certain area can be compared with the observations in the field. His examples show that this method demonstrates an altitude limit of vascular plants to be generally higher than those observed in the field.

As has been observed in the study areas (WA1-WA5) the number of species tend to drop rapidly near the altitude limit of vascular plants. When using a linear regression therefore to estimate the altitude limit of vascular plants would give a higher limit than that observed in the field.

8.3.4 Methods based on the cumulative altitude distribution of observations (type 2b method)

General remarks

To avoid the disadvantages of the estimation by the highest observation (type 1), other methods have been developed and tested. Several approaches have been studied by Geißdörfer (1994) from the stand point of a biometrician. One possibility is to use the function of the accumulated altitude distribution of observations as a base for the estimation of the altitude limit of vascular plants.

Database

Geißdörfer based his work on the 9348 observations from the study areas WA1-WA5 available at the end of 1992. This number is below the total of the 9626 observations because a number of the collected specimens from these areas has been identified and added to the database later. Furthermore, Geißdörfer (1994) uses a list of 183 species instead of 176 species in the present paper. The difference is due to the decision to unify some doubtful small-species into larger taxonomic units (see chapter 3) following Bay (1992) and Fredskild (1996 b).

Geißdörfer (1994) used observations instead of collected specimens. The number of observations (n = 9348) is about three times higher than the number of specimens (n = 3457). Plant lists from 621 sites are available instead of 373 sites, if herbarium specimens are used. The greater quantity of material gives a better picture of the altitude distribution of vascular plants.

One purpose of the study was also the comparison of the vertical distribution of vascular plants in areas with different climate and geology. To do this the observations in the five study areas WA1-WA5 were rearranged into nine groups called "provinces PA-PI" as is shown in Tab. 39.

Proposed method (type 2b)

All altitude records of observations/province are ordered from the lowest to the highest altitudes and accumulated. The resulting distribution of frequencies (y-axis) and altitudes (x-axis) describes a curve reaching the end, if the highest record has been added as the last record. The graphical presentation of this data is the starting point for fitting a suitable function to this distribution. This function is used to estimate the altitude limit of vascular plants based on the distribution of all observations within a "province".

The results of the statistical analysis have been published (Geißdörfer 1994, Urfer & Schwarzenbach 1995). These authors propose to estimate the altitude

Tab. 39. Comparison of study areas WA1-WA5 and provinces PA-PI.

WA	P	Region
WA1	PA	Southern Werner Bjerge, Pingo Pass, Schuchert Dal
WA1	PB	Eastern Stauning Alper: Syltoppene, Skeldal, Bersærkerbræ
WA2	PC	Northern Stauning Alper: Skjoldungebræ, Linné Gletscher, Sedgwick Gletscher
WA2	PD	Western Stauning Alper: Vikingebræ, Gullygletscher, Sefstrøm Gletscher
WA2	PE	Nathorst Land, Forsblad Fjord, Tærskeldal, Højedal
WA3	PF	South Andrée Land: Junctiondal, Månesletten, Benjamin Dal, Luciadal, Blåbærdal
WA3	PG	North Andrée Land: Moraenedal, Endeløs, Nøkkefossen, Kap Weber, Grejsdal
WA4	PH	Ole Rømer Land, Vibeke Sø, Promenadedal, Krumme Langsø, nunataks north of 74°N
WA5	PI	Kronprins Christian Land: Centrum Sø, Drømmebjerg, Ingolf Fjord, Hekla Sund

WA: Study area
P Province

limit by fitting a logistic function to the empirical distribution. They compare two methods:

- basing the estimation on all observations/province
- basing the estimation on only the 50% of the observations above the median of all observations/province.

With both methods it is possible to estimate the altitude limit of vascular plants for an assumed proportion of observations (e.g. 99.0%, 99.5% or 99.99%). The second approach gives the better results and is therefore recommended. These estimations can be compared with the observations in the field (Tab. 40).

Unfortunately, the estimations are biased by inconsistent altitude distribution of the sites within the three provinces (PE, PH, PI), which had to be excluded from the table. These are,

- Province PE: Too many plant lists from sites near the sea.
- Province PH: No plant list below the altitude of 200 m; two clusters at 200-250 m and at 450-500 m.
- Province PI: Too many plant lists from low levels below 100 m.

The values in the line "Estimation 1" are below the highest observations for the provinces PA-PC and PF; higher for the province PD, while in PG the estimated altitude limit of vascular plants just corresponds to the altitude of the highest site.

The estimations at the level of 99.9% ("Estimation 3") are higher than the observed maximum with a variation between 200 m (PC) and 530 m (PD).

The differences between the observed maxima and the estimations at the level of 99.5% ("Estimation 2") range between -180 m (PD) and +150 m (PC). The "Estimation 2" gives obviously a reasonable estimation for the altitude limit of vascular plants.

Tab. 40. Estimated altitude limit of vascular plants compared with the highest observations in the field (Provinces PA-PD, PF-PG)

	PA	PB	PC	PD	PF	PG
Highest sites	1430	1450	2200	1870	1620	1500
	1370	1250	1720	1690	1500	1440
	1360	920	1700	1670	1480	1400
	1270	730	1680	1670	1470	1350
	1220	725	1660	1540	1460	1330
Estimation 1 (99.0%)	1350	1300	1900	1900	1600	1500
Estimation 2 (99.5%)	1450	1500	2050	2050	1750	1650
Estimation 3 (99.9%)	1750	1950	2400	2400	1950	2000

If the altitudes above the estimations of the altitude limit of vascular plants (P = 99.0%) are counted, the following numbers result: Province PA (3), PB (1), PC (1), PD (0), PF (1), PG (1) just reaching the estimated value. The sum of 7 observations above the values of "Estimation 1" is just 1% of the 676 localities recorded in the database of all observations.

Geißdörfer (1994) has also used the logistic function for the estimation of the altitude limit of vascular plants based on plant-lists from different altitudes. It was planned to compare statistically the curves fitted to the accumulated vertical distribution of observations for the nine provinces PA-PI. However, the available samples did not fulfil the requirements.

8.3.5 Empirical approach for the estimation of the altitude limit of vascular plants (WA1-WA5)

If it is intended to use the altitude limit of vascular plants as a bioindication for the regional temperature conditions during the growing period, a combination of several parameters is proposed, which describe the altitude distribution of vascular plants in the uppermost altitude bands:

(a) Estimation by the method of the highest observation (= "max. observation").
(b) Estimation of the altitude passed by 1% of all observations of plants or by at least of the 10 highest observations.
(c) Estimation of the altitude passed by 5% of all species or by at least 5 species.
(d) Estimation of the altitude passed by 5% of all sites or by at least 5 sites.

Herbarium specimens

The results obtained by these four methods are summarised based on the 3457 herbarium specimens (Tab. 41).

The estimation (a) gives the highest, the estimation (d) the lowest values for all five study areas. The estimations (b) and (c) are in between of the two extremes. The differences between the lowest value and the maximum in each row vary between 180 m (WA1, WA2) and 230 m (WA5). If the five study areas are compared the following results are obtained (Tab. 41):

- WA2 has the highest altitudes, WA5 the lowest altitudes within all four columns. The differences (WA2-WA5) in the five columns are explained by the fact, that WA5 is situated 8° further north and is not far from the outer coast.
- The altitudes of WA2 are considerably higher than the corresponding figures in WA1. These differences are explained by a strong gradient of the

Tab. 41. Estimation of the altitude limit of vascular plants WA1-WA5 (based on herbarium specimens only)

Study area	max. obs. (a)	1% of obs. (b)	5% of species (c)	5% of localities (d)
WA1	1220 m	1200 m	1200 m	1040 m
WA2	1720 m	1680 m	1700 m	1540 m
WA3	1500 m	1330 m	1350 m	1285 m
WA4	1470 m	1370 m	1470 m	1270 m
WA5	900 m	690 m	700 m	670 m

summer temperature between the more oceanic conditions of WA1 and the continentality of WA2. If a gradient of 0.5°C/100 m is assumed the observed difference in the altitude limit of vascular plants corresponds with an estimated difference of 2.5 C.

- The altitude limits of vascular plants in WA3 and WA4 are similar. The influence of the higher latitude of WA4 (74° instead of 73°) might be compensated by the more continental conditions of WA4.

All observations

The results of the estimations are based on all 9626 field observations instead of the 3457 herbarium specimens are shown in Tab. 42.

- The estimation of "highest observation" (a) gives the highest, the "highest 5% of the localities" (d) gives the lowest values for all five study areas. The estimations by "1% of observations" (b) and "5% of species" (c) are in between of the two extremes.
- The differences between the lowest value and the maximum in each row vary between 125 m (WA5) and 580 m (WA2).
- WA2 has the highest, WA5 the lowest values within all four columns. The differences (WA2-WA5) are 1245 m, 790 m, 915 m and 790 m. The unusually high difference in column (a) is explained by the outlying observation of *Saxifraga cernua* on the summit crest of Elisabeth Bjerg. The average of the

Tab. 42. Estimation of the altitude limit of vascular plants (based on all 9626 field observations)

Study area	max. obs. (a)	1% of obs. (b)	5% of species (c)	5% of localities (d)
WA1	1470 m	1220 m	1370 m	1040 m
WA2	2200 m	1700 m	1870 m	1620 m
WA3	1620 m	1500 m	1620 m	1400 m
WA4	1470 m	1350 m	1470 m	1270 m
WA5	955 m	910 m	955 m	830 m

four differences is greater than that calculated from Tab. 41 "herbarium specimens only" (935 m vs. 920 m).

- The altitudes of WA2 are considerably higher than the corresponding figures in WA1.
- The altitude limits of vascular plants of WA3 and WA4 differ by 145 m based on the averages of all four estimations (1535 m in WA3 and 1390 m in WA4).

8.3.6 A simple and fast method for the estimation of the altitude limit of vascular plants

Based on the results of the previous chapter a simple and fast method is proposed. The observations are arranged according to altitude beginning with the highest value. The sequence is followed until the altitude of *the fifth highest species* has been reached. This value is then used as an estimation of the altitude limit of vascular plants (Tab. 43 and 44).

The altitude of the five highest species show the altitude range of about 4-6% of the number of species found in WA1-WA5 (see Tab. 9). A high range would be expected, if outliers occur. The method shows therefore the reliability of the estimation of the altitude limit. It is proposed to give the altitude range of the five highest species instead of the altitude of the only the highest species.

The altitude range between the first and the fifth value within each column

Tab. 43. Estimation of the altitude limit (meters) of vascular plants based on the five highest species (herbarium specimens only)

WA1	WA2	WA3	WA4	WA5
1220	1720	1500	1470	900
1200	1720	1500	1470	810
1200	1720	1500	1470	770
1200	1720	1440	1470	770
1200	1720	1400	1470	770

Tab. 44. Estimation of the altitude limit (meters) of vascular plants based on the five highest species (9626 observations)

WA1	WA2	WA3	WA4	WA5
1450	2200	1620	1470	955
1450	2050	1620	1470	955
1450	2050	1620	1470	955
1450	1970	1620	1470	955
1430	1870	1620	1470	955

(Tab. 43) varies between the study areas: WA1 (20 m); WA2 (350 m); WA3 (0 m); WA4 (0 m); WA5 (0 m). A range of "0 m" means that at least five species have been collected at the highest recorded site. A wide range as in WA2 is the result of a species being found at scattered isolated sites with different altitudes.

If the estimation is based on all 9626 observations (Tab. 44) the range between the altitude limits of the five highest highest species is 330 m in the column of WA2.

8.3.7 Conclusions

When observations are available only from altitudes just below and above the assumed altitude limit of vascular plants the estimation is best done using the method of the five highest species (8.3.6). This approach also includes the highest observation of a vascular plant in the study area [method (a) in 8.3.5]. If the maxima for each of the five species are noted, the range of these altitudes gives a reasonable estimation of the altitude level of the highest band where vascular plants are found. This method has the advantage that neither the total number of species, nor the pattern of the altitude distribution of species in middle or lower altitude require to be known.

When field observations are available from all altitude levels, several methods can be used. If observations were available from sea level to the highest altitude, then it is recommended to use first the methods outlined here (b), (c) or (d) of 8.3.5. These could then be compared with with "Estimation 2, 99.5%" calculated by the method described in 8.3.4, Tab. 40.

The method proposed by Halliday (8.3.3) calculating the linear regression "rate of decrease number of species/100 m" is a useful approach. However field observations suggest, that the calculated value for the altitude limit of vascular plants by this method is in general too high.

9. Comparison of the altitude distribution of different areas in Greenland

9.1 Methods and data

The pattern of vertical distribution within a defined area can be described by the descending order of the highest records of the species. This analysis of the altitude zonation is based on all 9626 field observations and not on the 3457 collected specimens. This greater quantity of material gives a better representation of the vertical distribution.

However, two limitations should be noted to avoid misinterpretations:

- The study includes only the most common species represented in the database by at least 50 observations .
- The species of the genera *Melandrium, Potentilla* and *Taraxacum* difficult to identify in the field have been omitted. The genus *Draba* is only represented by *D. subcapitata.* The species *Poa alpina s.l.* has also been omitted.

The 42 commonest species selected for this analysis of the altitude zonation are listed in Tab. 45. The highest record of each species in each of the five study areas has been given. The last column ,Av.' shows the average of the values/row. The species are arranged in descending order (columns ,Av.'). Missing observations are symbolised by the sign ,-'. Within the list the following five species are marked: *Salix arctica, Cassiope tetragona, Dryas octopetala s.l., Vaccinium uliginiosum* and *Betula nana.* These species show the relative altitude distribution of the two ,tree' species and three dominant species of heath to illustrate the altitude zonation.

9.2 Altitude zonation and description of the 42 commonest plants

The three continental study areas WA2 – WA4 have the highest average altitudes with a range between 1516 m (WA2) and 1161 m (WA4). The value of WA4 (74°N) is 355 m lower than the average in WA2 (72°N) indicating a gradient of about 175 m per one degree of latitude. The difference of 874 m between WA2 (72°N) and WA5 (80°N) corresponds to a remarkable decrease of the mean temperature during the growing season estimated as c. 4.4°C. The

Tab. 45. Highest altitudes of 42 commonest species (50 or more observations of each species) in WA1-WA5

Species	WA1	WA2	WA3	WA4	WA5	Av.
Saxifraga cernua	1370	2200	1620	1470	955	1523
Papaver radicatum s.l.	1430	1870	1620	1470	955	1469
Campanula uniflora	1200	1720	1470	1470	*	1465
Cerastium arcticum s.l.	1270	1870	1620	1470	955	1437
Saxifraga caespitosa s.l.	1250	1870	1620	1470	955	1433
Saxifraga oppositifolia	1450	2050	1310	1470	830	1422
Luzula confusa	1200	1870	1620	1350	955	1399
Saxifraga nivalis	1220	1870	1480	1350	955	1375
Carex nardina	1450	1720	1500	1270	830	1354
Poa abbreviata	1220	1500	1620	1470	955	1353
Salix arctica	**1360**	**2050**	**1400**	**1180**	**750**	**1348**
Minuartia rubella	1200	1670	1310	1350	955	1297
Poa arctica	1200	1700	1500	1350	710	1292
Trisetum spicatum	1200	1720	1500	1180	670	1254
Chamaenerion latifolium	1140	1700	1460	1170	770	1248
Draba subcapitata	1200	1670	1200	1350	810	1246
Poa glauca	1100	1870	1260	1350	650	1246
Cystopteris fragilis s.l.	1450	1460	1480	1470	330	1238
Carex misandra	1200	1670	1250	1180	670	1194
Luzula arctica	1000	1360	1150	1300	750	1112
Polygonum viviparum	1200	1400	1500	960	420	1096
Cassiope tetragona	**1090**	**1400**	**1320**	**900**	**730**	**1088**
Carex rupestris	770	1460	1250	1180	670	1066
Juncus biglumis	1200	900	1310	1180	700	1058
Carex capillaris s.l.	650	1420	1210	930	–	1053
Dryas octopetala s.l.	**1010**	**960**	**1300**	**930**	**–**	**1050**
Braya purpurascens	1090	*	1300	1170	600	1040
Silene acaulis	1000	1700	1040	930	470	1028
Oxyria digyna	1200	1010	1310	790	710	1004
Eriophorum triste	1120	1360	1140	900	470	998
Phippsia algida s.l.	620	900	1380	1200	810	982
Kobresia myosuroides	800	1420	1200	1180	200	960
Calamagrostis purpurascens	690	1500	1285	960	330	953
Pedicularis hirsuta	1040	945	1180	900	500	913
Pyrola grandiflora	650	1450	985	540	–	906
Vaccinium uliginosum	**770**	**1420**	**1250**	**780**	**200**	**884**
Pedicularis flammea	770	1230	1100	900	200	840
Carex scirpoidea	770	1110	650	450	–	745
Arctagrostis latifolia	1000	900	720	780	310	742
Carex bigelowii s.l.	1000	1030	1080	450	150	742
Betula nana	**760**	**870**	**700**	**570**	**–**	**725**
Saxifraga aizoides	1000	750	870	640	240	700
Number of species	42	41	42	42	36	–
Average	1121	1516	1330	1161	642	–
Difference compared with WA2	395	0	186	355	874	–

Abbreviations:
WA1-WA5 study areas
Av. average altitude
* a single observation at a very low altitude has been omitted therefore

difference of 395 m between WA2 and WA1 is explained by the effect of the more coastal climate in WA1.

In Tab. 45 the species are arranged in descending order of the averages in the column "Av.". The following results are of some importance:

- The average of the altitude limits is above 1000 m for 29 of the 42 species (69%).
- The species of the vegetation type „fell-field" are found in the uppermost part of the table. *Cassiope tetragona* ranges in the middle with an average of 1088 m. Species of mossy heaths such as *Eriophorum triste* (998 m), *Carex bigelowii s.l.* (742 m), *Arctagrostis latifolia* (742 m) and *Saxifraga aizoides* (700 m) are found only in the lower part of Tab. 45.
- The altitudes within the five columns WA1-WA5 also decrease from top to bottom. A comparison shows some inconsistences between the five lists which can be explained partially by checking the field notes: *Saxifraga cernua* was observed at the extreme altitude of 2200 m (WA2, summit crest of Elisabeth Bjerg).
- *Saxifraga oppositifolia* and *Salix arctica* were observed at 2050 m (WA2, Schaffhauserdalen, sheltered site on a steep slope).
- *Cystopteris fragilis* is usually found in sheltered niches below big blocks or in cracks in rocks. Therefore, the altitude distribution is often scattered, which explains the only observation at 330 m in WA5.
- *Polygonum viviparum*: The remarkable difference between the highest sites in WA3 (1500 m) and WA4 (960 m) cannot be explained.
- *Cassiope tetragona* was found in WA4 only at lower altitudes, because the soils in the nunataks are too dry.
- *Carex rupestris*: Sites with favourable conditions are rare in WA1 at higher altitudes.
- *Juncus biglumis* (WA2, 900 m) and *Oxyria digyna* (WA2, 1010 m): conditions in the interior of the Stauning Alper are not suitable.
- *Carex capillaris s.l.* has been found in WA1 only once (Pingo Pass, 650 m).
- *Dryas octopetala s.l.* has a low altitude limit on the cystalline rocks of the Stauning Alper (960 m). The species reaches much higher altitudes on the limestone/dolomite sediments in WA3 (1300 m).
- *Silene acaulis* was growing in WA2 at an unusual high site (1700 m) in the interior of the Stauning Alper at the foot of a sunny rock-wall.
- *Phippsia algida s.l.* has been found in WA1 (620 m) and in WA2 (900 m) only at low altitudes.
- *Kobresia myosuroides* prefers the continental climate of the interior fjords. This species has been observed in WA1 (800 m) and WA5 (200 m) only at low altitudes.
- *Calamagrostis purpurascens* is a species typical for the ‚steppe'. This species

has been found in WA1 (690 m) and in WA5 (330 m) only at a rather low altitudes.

- *Pedicularis hirsuta* was rather rare in the interior of the Stauning Alper reaching in WA2 only an altitude of 945 m.
- *Pyrola grandiflora* was found at the high altitude of 1450 m in the western Stauning Alper.
- *Vaccinium uliginosum* var. *microphyllum* reaches fairly high altitudes in the continental parts of WA2 (1420 m) and WA3 (1250 m), but is restricted to the low ground in WA5 (200 m).
- *Carex bigelowii s.l.*, *Arctagrostis latifolia* and *Saxifraga aizoides* have been collected at an unusually high site at 1000 m in WA1. The same species have been collected only at low levels in WA2 (1030 m) where the conditions are unfavourable for wet heaths and mires.

9.3 Comparison between the study areas WA1-WA5, and comparisons and with other areas in Greenland

9.3.1 Data

The method of altitude zonation of the 42 commonest species has been used to compare the altitude distribution in the study areas WA1-WA5 inter se and with the pattern in other areas of Greenland.

Gelting (1934: 241) has published a table showing the altitude distribution of species on the southern coast of Clavering Ø, based on his observations from 1931/32. He compares his results with data from Scoresby Sund (observations by Hartz 1891/92), from West Greenland (Porsild 1926, observations by A. E. Porsild 1924), and from Southwest Greenland (observations by Rosenvinge 1892 and by A. E. Porsild 1924). Holmen (observations 1957) has indicated the highest altitude records for most of the species found in Peary Land (82°N). His data are of special interest for a comparison with the observations in Kronprins Christian Land (WA5, 80°N). Finally, Schwarzenbach's observations (unpublished) in Gåse Land (70°N, 1994), and in North Peary Land (83°N, 1995) are used for comparisons. The results are listed in Tab. 46.

These six additional lists are combined with data from WA1-WA5 to form three groups based on latitude:

- Group G1, 70° – 71°N: Scoresby Sund (I, II), Southern Werner Bjerge (WA1) and West Greenland (VI)
- Group G2, 72° – 74°N: Clavering Ø (III) with the two study areas WA3 and WA4
- Group G3, 80° – 83°N, WA5 and Peary Land (IV, V)

Tab. 46. Highest observations of 42 selected species (50 or more observations) in six other areas of Greenland

Species	I	II	III	IV	V	VI	AWA
Saxifraga cernua	1255	1500	1200	1100	660	940	1523
Papaver radicatum s.l.	1570	1500	1250	1000	970	1420	1469
Campanula uniflora	1570	1050	1000	50	–	940	1465
Cerastium arcticum s.l.	1570	1500	1250	1100	990	1255	1437
Saxifraga caespitosa s.l.	1570	1500	1225	1100	710	1420	1433
Saxifraga oppositifolia	1570	1500	1200	1100	940	770	1422
Luzula confusa	1570	1170	1225	900	910	960	1399
Saxifraga nivalis	1570	1500	1225	600	670	930	1375
Carex nardina	1570	1170	1200	–	500	930	1354
Poa abbreviata	–	–	1125	1100	930	–	1353
Salix arctica	**1570**	**1170**	**1200**	**1100**	**660**	**770**	**1348**
Minuartia rubella	1320	1000	1000	1000	910	1420	1297
Poa arctica	1570	1000	1200	1000	660	–	1292
Trisetum spicatum	1255	1500	1100	600	380	–	1254
Chamaenerion latifolium	1255	990	900	150	480	770	1248
Draba subcapitata	1570	–	1225	1000	760	–	1246
Poa glauca	–	1170	1125	1000	600	1255	1246
Cystopteris fragilis s.l.	1255	1000	900	500	–	500	1238
Carex misandra	1100	1170	1100	–	660	910	1194
Luzula arctica	–	1160	1100	900	660	960	1112
Polygonum viviparum	1255	1170	1200	600	580	960	1096
Cassiope tetragona	**1570**	**1170**	**1000**	**–**	**540**	**910**	**1088**
Carex rupestris	940	1000	1000	50	–	770	1066
Juncus biglumis	785	250	1200	500	510	770	1058
Carex capillaris s.l.	–	380	900	–	–	–	1053
Dryas octopetala s.l.	**1500**	**1160**	**1125**	**–**	**–**	**930**	**1050**
Braya purpurascens	–	–	1100	200	480	770	1040
Silene acaulis	1570	1170	1200	200	–	1420	1028
Oxyria digyna	940	1150	900	600	460	940	1004
Eriophorum triste	750	560	1000	500	370	785	998
Phippsia algida s.l.	785	–	1225	1000	660	940	982
Kobresia myosuroides	626	1060	1000	600	500	500	960
Calamagrostis purpurascens	–	950	1000	–	230	785	953
Pedicularis hirsuta	1320	1120	900	200	540	770	913
Pyrola grandiflora	1320	1000	500	–	–	930	906
Vaccinium uliginosum	**1320**	**1010**	**900**	**150**	**–**	**770**	**884**
Pedicularis flammea	940	460	800	–	–	–	840
Carex scirpoidea	785	850	600	–	–	770	745
Arctagrostis latifolia	1320	560	900	500	560	300	742
Carex bigelowii s.l.	625	1120	1125	700	–	1320	742
Betula nana	**940**	**860**	**700**	**–**	**–**	**–**	**725**
Saxifraga aizoides	125	–	600	–	–	–	700
Average	1226	1080	1039	681	637	900	1144
Number of species	36	36	42	31	29	34	42

Abbreviations:

AWA		average of all five study areas WA1-WA5 for comparison	
I	70°N	East Greenland	Scoresby Sund (Hartz 1891/92)
II	70°N	East Greenland	Gåse Land, Scoresby Sund (Schwarzenbach 1994)
III	74°N	Northeast Greenland	Clavering Ø (Gelting 1933)
IV	82°N	North Greenland	Peary Land (Holmen 1957)
V	83°N	North Greenland	North Peary Land (Schwarzenbach 1995)
VI	70°-71 N	West Greenland	West Greenland (A. E. Porsild 1924)

9.3.2 Group G1: Four areas in East and West Greenland, 70°-71°N

The averages of the areas I, II, WA1 and VI are calculated only for the species present in all four areas, and differ therefore slightly from the values in Tab 45 and 46. The following averages have been calculated:

- Area I, Scoresby Sund, 1242 m
- Area II, Gåseland, 1074 m
- Study area WA1, Southern Werner Bjerge, Syltoppene, 1086 m
- Area VI, West Greenland, 897 m

The averages of the three areas in East Greenland (I, Scoresby Sund; II, Gåseland; WA1, Southern Werner Bjerge, Syltoppene) vary from 1074 m to 1242 m. They are higher than the average for the area VI in West Greenland with 897 m. As the three areas in East Greenland belong to the continental belt of the interior fjords, they could have higher temperatures during the summer than the region in West Greenland.

The averages of the two areas II (Gåse Land, 1074 m) and WA1 (Southern Werner Bjerge, 1086 m) have similar values. The average for area I (Scoresby Sund , 1242 m) is higher than the average for Gåseland (1074 m). This result is surprising. Hartz (area I) based his list on records from south-facing slopes at the eastern coast of Gåseland, while Schwarzenbach (area II) has worked not far away in western Gåseland. The observed difference could be explained by the exposure: Most of the sites in area II (lower Hjørnedal, Sæfaxa Sø) have unfavourable exposures.

9.3.3 Group G2: Three areas in Northeast Greenland, 72°-74°N

The averages are calculated only for the species present in all three areas (WA3, WA4, III). The averages of the study area WA3 (1288 m) and of WA4 (1104 m) are above the average for area III, Clavering Ø (1039 m). The differences are explained by the higher latitude of WA4 and area III on one side, and by the climatic influence of the outer coast on the other.

9.3.4 Group G3: Three areas in North Greenland, 80°-83°N

The averages are calculated only for the species present in all three region. They are similar as shown in the following list:

- Study area WA5, Kronprins Christians Land, 660 m
- Area V, Peary Land, 688 m
- Area VI, North Peary Land, 615 m

10. Altitude distribution and diversity

10.1 Previous studies

The diversity of the mountain vegetation of East and Northeast Greenland has been subject of several studies and is part of a joint research program with Prof. Dr. W. Urfer, Department of Statistics, University of Dortmund (Germany) and the author. The same data referred to in chapter 9 have been used for the analysis of diversity.

The following questions have been studied in detail:

- Which methods are best suited to analyse the diversity of the mountain vegetation in a region with latitude and altitude limits of vascular plants?
- What degree of diversity of the vegetation is observed in the mountains of East and Northeast Greenland where the plant cover has developed under natural conditions since the time of the latest glaciation?
- Does the observed species-diversity reflect a steady state of a self-regulating dynamic process? Is it possible to explain which factors influence the long-term development of the vegetation?

10.2 Measurement of diversity

The diversity of a plant-community can be described and measured by different methods. A well-known approach is the calculation of the "Shannon/Wiener-Index of diversity" for the so-called "species/abundance-distribution of a sample". This index is based on a table showing the list of species found in the sample and the relative abundance of each species. As a rule the species are arranged in descending order according to their abundance, and calculated in percentages of the total number of individuals in the sample.

Many publications deal with the definitions and the estimations of "diversity indices" (references in Kaiser 1994, Urfer & Schwarzenbach 1995, Schwarzenbach et al. 1996 b).

10.3 Results

The first analysis of the data from Greenland has been done by Kaiser (1994). She has discussed the formal aspects of several indices proposed in the literature for the description and the measurement of diversity (Simpson 1949, Mac Arthur 1972, Piélou 1975, Kempton & Wedderborn 1978, Kempton 1979,

Patile & Taillie 1982, Washington 1984, Stugren 1986, Magurran 1988, Wilson & Peter 1988, Krebs 1989). Based on the 9348 observations from nine areas (called "provinces PA-PI", see chapter 9) she has calculated the Shannon-index (Sh-index), the Simpson-index (Si-index) and the Q-statistics (Kempton & Wedderborn 1978). In addition she has compared the resulting Sh-indices with the theoretical maximum to be expected in the case of equal distribution of all species. The quotient between the empirical and the theoretical value measures the degree of homogeneity of the sample and is called "measurement of homogeneity for the Sh-index [(V-)-Sh]".

Kaiser (1994) has compared the Sh-index, the (V)-Sh, the Si-index and the Q-statistics for the nine provinces and has studied the question, how these measurements change in relation to altitude above sea-level. Urfer & Schwarzenbach (1995) published the first results of an analysis of the diversity of the mountain vegetation in East and Northeast Greenland.

A short introduction describes and explains the methods for the measurement of diversity based on the number of observations of each species recorded within a province. The indices calculated for this so-called "species/presence-distribution" offer a good basis for the analysis of diversity of the rather open plant-cover near the latitudinal and altitude vegetation-line. The following results are of interest (Tab. 47-49):

- The values obtained for the Sh-index, the Si-index and the Q-statistics show a similar diversity of the vegetation in all nine provinces. It seems that the vegetation of the arctic mountains has established a similar equilibrium

Tab. 47. Shannon-Index: Estimations

PA	PB	PC	PD	PE	PF	PG	PH	PI	PA-PI
4.27	4.03	4.07	4.16	4.19	4.33	4.42	4.42	4.08	4.51

Explanation:
PA, PB, ...PI provinces
PA-PI all observations of provinces PA to PI

Tab. 48. Simpson-Index: Estimations

PA	PB	PC	PD	PE	PF	PG	PH	PI	PA-PI
0.981	0.976	0.979	0.982	0.980	0.984	0.986	0.986	0.978	0.984

Explanation: (see Tab. 47)

Tab. 49. Q-Statistics: Estimations

PA	PB	PC	PD	PE	PF	PG	PH	PI	PA-PI
38.30	30.58	26.59	27.91	28.74	35.70	38.72	34.73	24.65	39.03

Explanation (see Tab. 47)

between the species, independent of the diverging sets of species in the nine provinces. A few relatively small deviations from the common average could be explained by peculiarities of the data-sets or by ecological differences between the provinces.

- The vegetation is rather homogeneous in all provinces with values for the Si-index near the theoretical maximum.
- The Sh-index decreases with increasing altitude while the index of homogeneity increases and reaches nearly the theoretical maximum (Fig. 40, 41).

Generally, the species/presence-distribution becomes more homogeneous whenever the vegetation reaches an ecological limit caused e.g. by drought, dessication or salinity. (Schwarzenbach et al. 1996 b). There is a simple explanation: A high index of homogeneity is reached, when a few species have the same abundance within the analysed samples. This situation has been confirmed for many sites near the altitude limit of vascular plants, where a few "high altitude species" have equal chances to reach and to occupy the highest niches of survival.

On the other hand, the homogeneity decreases form high to low levels, as it is shown by Fig. 41. This trend is explained by a higher proportion of rare and often highly specialised species at low altitudes. Therefore, the Sh-index increases, while the measurement of heterogeneity (V)-Sh indicates a lower degree of homogeneity.

The diversity of vegetation may be influenced by the competition between

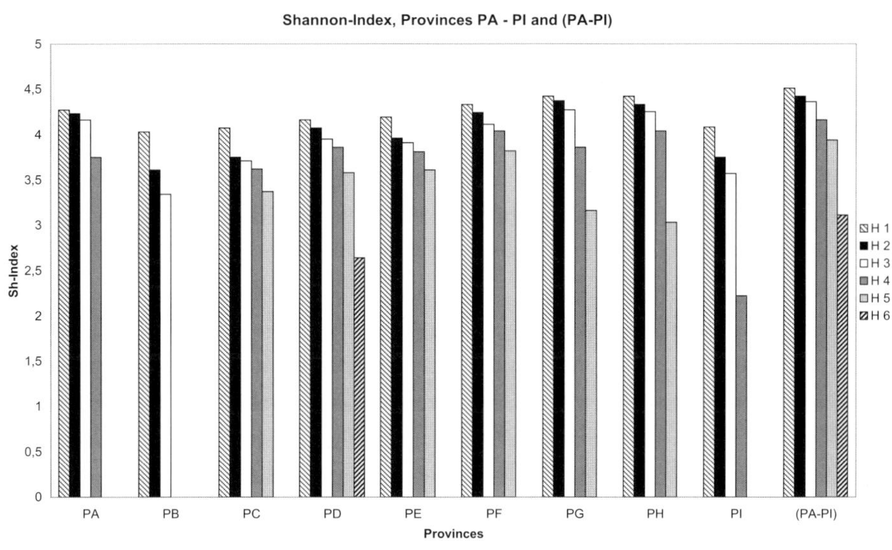

Fig. 40. Shannon Index, provinces PA-PI and (PA-PI). H 1: 0-299 m, H 2: 300-599 m, H 3: 600-899 m, H 4: 900-1199 m, H 5: 1200-1499 m, H 6: 1500-1799 m.

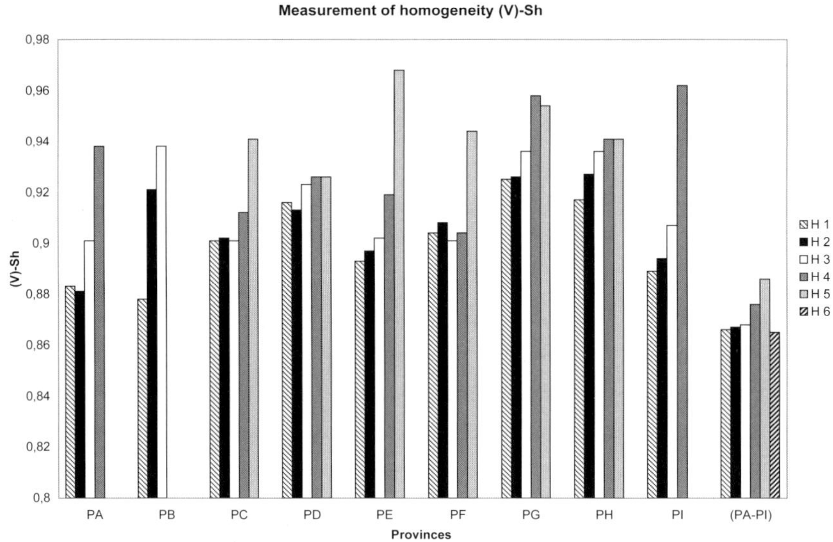

Fig. 41. Measurement of homogeneity (V)-SH, provinces PA-PI. Explanations see fig. 40.

species. However, as sites with homogenous ecological conditions and with a closed cover of vegetation are rare in the mountains of Greenland, the present material is not sufficient for an analysis of this aspect.

11. Climatic aspects and altitude distribution

11.1 An old problem: "Changes of climate and vegetation"

The relationship between changes in climate and the corresponding changes in the vegetation have been studied and discussed from various aspects for more than a century. Increased interest in the subject has been shown by scientists in recent years because of the possible influence of the "greenhouse-effect" on climate.

The effect of climatic changes on Greenland vegetation has mainly been concentrated on the Holocene Period, and several questions have been discussed in the literature:

- Did the Greenland flora develop before the Quaternary period and did it survive – at least with some elements – through repeated glaciations?
- Which type of vegetation existed in Greenland during interglacial or interstadial periods?
- Have species from an interglacial or interstadial vegetation survived? If so, which areas free of ice have been refugia for these elements?
- Have species survived on nunataks or in unglaciated areas of the lowland?
- How has the geographic and the topographical distribution of vascular plants changed in Greenland during the climatic changes of the Pleistocene?

Different methods have been used to answer these questions:

- Palynological and ice core analysis have contributed a lot of interesting and important results in the palaeobotanical investigations in Greenland (e. g. Funder & Abrahamsen 1988, Fredskild 1973, Fredskild 1996 c).
- Mapping local advances of glaciers in former times (e. g. glacial deposits, polished rocks) and mapping raised beaches, to provide information about the changes in local glaciation as an indication of climatic change (Gelting 1934).
- Studies on quaternary geology of the ice-free shelves and adjacents shelves of Greenland have provided valuable information on the history of glaciation in Greenland (Funder & Hjort 1973, Funder 1989 a, b).
- The analysis of the present geographic distribution of species in Greenland as well as over the whole Arctic may be a promising method. This approach profits from the better knowledge gained by several regional monographs

(for example Böcher 1933, 1938, 1951 b, 1956, 1959, 1963), Bay (1983, 1992), Feilberg (1984), Fredskild (1973, 1996 b), and regional contributions (for example Bay & Boertmann 1989), Bay & Fredskild (1989, 1990), Graham & Grimm (1990), Huntley (1991).
- The altitude distribution of species within an area and the analysis of sites at high altitude is an additional approach as it is shown later in this chapter.

11.2 Previous studies

Gelting (1934) discussed many aspects of the subject based on his observations from Clavering Ø and adjacent areas. He has summarised the state of knowledge at that time about unglaciated areas in Greenland and presents a list of authors who have explored this subject. He mentions Kolderup (1879) as the oldest reference.

Eberlin (1888) made the following statement regarding the flora of Southeast Greenland (cited by Gelting [1934] as translation from the Danish text): "As a strong subjective impression I shall claim that the greater number of the flowering plants in Danish East Greenland are the relicts of a South Greenland Ice-Age flora highly decimated by the present "wet" climate".

Later there was a sharp controversy between Warming (1888) and Nathorst (1890, 1891) on this topic (summarised by Fredskild 1991).

Nowadays there is general agreement among scientists that there were unglaciated areas in Greenland at least during the latest glaciation of the Ice Age (e. g. Funder 1989 a, b). Besides flat areas of low or middle altitude there are numerous sites free of ice in the mountains of alpine character as in the Stauning Alper (72°N). Niches occupied by vascular plants even at altitudes above 1600 m suggest that at least some species could have survived over a long period with temperatures much lower than today.

Funder (1989 a, b) has summarised the Quaternary succession on Jameson Land, and its proposed correlation with North European chronostratigraphy and marine oxygen isotope stratigraphy.

11.3 Influence of climate and climatic changes on the altitude distribution of plants

11.3.1 Approach and analysis

The influence of climate and climatic changes on the vegetation can be elucidated by analysis of the present geographic and altitude distribution of vascular plants in the mountains of East and Northeast Greenland. The analysis requires several steps.

A. Firstly, the present geographic and altitude distribution of species has been described in detail.
B. Secondly, the general pattern of distribution has been considered with reference to macro-climatic gradients of summer temperatures. For example:
   B 1) decrease in numbers of species with increasing northern latitude, and with increasing altitude
   B 2) increase of altitude limits caused by mountain mass and increasing continentality.
C. Thirdly, the sites are listed when rare or unexpected species were collected there. These sites – called "niches" in this paper – have been mapped.
D. The fourth step is an attempt to explain the occurrence of the observed species in these "niches". Several hypotheses should be discussed:
   D 1) Is the irregular pattern of altitude distribution near the sea caused by an inversion of temperature during the growth season?
   D 2) Is the geographic pattern of altitude and north-south distribution influenced by the presence of glaciers?
   D 3) Is the pattern caused by special ecological conditions of the habitat?
   D 4) Is the observed species represented by a taxonomic, cytological or ecological sub-unit adapted to the special conditions of this habitat?
   D 5) Is the niche a possible pioneer-site of an immigrating species?
   D 6) Is the niche a possible refugium, which has been reached by a population of a southern species during a period with a warmer climate and, where the species might have survived the following advances of the ice and survived in these places during the latest advances of the ice?
   D 7) Have populations of northern species migrating during an earlier advance of the ice reached this isolated niche at high altitude during a later period when the climate was warmer?
   D 8) Can a disjunct pattern of altitude and geographic distribution be explained by other reasons?
   D 9) Are there refugial niches of northern species ?
   D 10) Are there disjunct patterns of altitude and geographic distribution?

11.3.2 List of niches

Searching through the files of all 176 species within the 26 regions (Tab. 10), a list of niches in isolated places and at unusually high altitudes has been extracted. For each of the 18 niches (N is used to denote niche and is numbered from 1-18) a short comment and a summary of the most interesting collections is given. These are shown within the study areas WA1-WA5.

WA1: N 1, Southern Werner Bjerge, mountains south of Randspids, 1200 m

A field of flat stones irrigated by melt-water from a late snow-bed was found on a terrace sloping slightly in south-eastern direction. A large number of species was collected in the centre and along the edges of this wet area. An interesting group of plants from snow-patches and mires have been found including *Polygonum viviparum, Draba alpina s.l., D. gredinii, Juncus biglumis, Carex maritima, Festuca hyperborea.*

WA1: N 2, Southern Werner Bjerge, north of Pingo Pass, 1000 m

This site is an active polygon-field on the south-facing slope of the mountain alongside a rivulet, where many species of mires, wet heaths, and wet sands have reached a place of unusually high altitude. Plants found include *Equisetum variegatum, Cochlearia groenlandica, Pedicularis hirsuta, Saxifraga aizoides, Eriophorum triste, Carex bigelowii s.l., Arctagrostis latifolia, Colpodium vahlianum.*

WA1: N 3, Southern Werner Bjerg, west of Biskop Alf Dal, 770 m

There are several early melting snow-patches not far away from the old western side-moraine of Biskop Alf Dal. The sheltered sites along the sunny and rather steep slope offer excellent conditions for the development of a herb-slope with southern species. Plants found include *Arabis alpina, Arnica angustifolia, Botrychium lunaria, Gentiana tenella, Poa alpina s.l , Potentilla crantzii, Sibbaldia procumbens, Taraxacum brachyceras, Thalictrum alpinum.*

WA2: N 4, Stauning Alper, Sefstrøm Gletscher, 960-1260 m (Fig. 42)

Many rare species are found at sheltered spots with southern or south-western exposure following the Sefstrøm Gletscher behind the eastern moraines. These sites form a chain of niches at different altitudes, and are set out below.

1260 m *Luzula spicata, Silene acaulis*

1230 m *Pedicularis flammea, Euphrasia frigida, Arnica angustifolia, Taraxacum arcticum, Calamagrostis purpurascens, Hierochloë alpina*

1200 m *Ranunculus pygmaeus, R. affinis s.l., Draba fladnizensis, D. glabella, Campanula uniflora, C. gieseckiana, Antennaria canescens, Carex rupestris,*

1030 m *Huperzia selago ssp. arctica, Salix herbacea, Sibbaldia procumbens, Minuartia biflora, Harrimanella hypnoides, Erigeron humilis, Taraxacum brachyceras, Tofieldia pusilla*

1010 m *Antennaria porsildii*

960 m *Dryas octopetala s.l., Pyrola grandiflora, Empetrum nigrum* ssp. *hermaphroditum, Vaccinium uliginosum* ssp. *microphyllum, Carex capillaris s.l, C. scirpoidea*

Fig. 42. Plants found at the foot of a rock (WA2, 1954). Stauning Alper, niche N 4, Sefstrøms Gletscher.

WA2: N 5, Stauning Alper, northern side of Vikingebræ, 360-870 m (Fig. 33, 34, 36, 43)

The steep slope at the northern side of the Vikingebræ offers sites with sheltered and sunny conditions for southern species. There is a chain of niches at different altitudes.

870 m *Betula nana, Pedicularis flammea, Pyrola grandiflora, Carex scripoidea*
850 m *Carex bigelowii*
810 m *Potentilla crantzii, Sibbaldia procumbens, Draba glabella, Viscaria alpina* (Fig. 36), *Euphrasia frigida, Taraxacum brachyceras, Festuca rubra s.l.*
360 m *Daba aurea* (Fig. 34), *Arabis holboellii* (Fig. 33)

A2: N 6, Stauning Alper, Gullygletscher, 1230-1460 m

Along both sides of Gullygletscher there are two high sites (1670 m and 1460 m) with species belonging to ADT A1 and B1. Unusually there are also the three species listed below.

1460 m *Silene acaulis*
1230 m *Chamaenerion latifolium, Antennaria canescens*

WA2: N 7, Stauning Alper, Skjoldungebræ, 1240-1700 m

Seven species of ADT A1 and B1 have been found at 1700-1720 m on south-facing rocks of the innermost part of Skjoldungebræ. A few niches of high al-

Fig. 43. Stauning Alper, niche N 5, northern side of Vikingebræ (WA2, 1954).

titude are also found on the western side and in the valley of an eastern side-glacier.

1700 m *Chamaenerion latifolium, Silene acaulis*
1660 m *Luzula spicata,*
1320 m *Campanula gieseckiana s.l., Antennaria canescens, Carex atrofusca*
1240 m *Huperzia selago*

WA2: N 8, Stauning Alper, Linné Gletscher, 1030-1335 m (Fig. 44)

There are several small niches with southern or rare species at sheltered and sunny sites in the rocks and behind the lateral moraines of the main glacier and the tributary glaciers.

1335 m *Campanula gieseckiana s.l., Antennaria canescens, Luzula spicata*
1300 m *Carex glacialis*
1150 m *Draba glabella, Erigeron compositus*
1120 m *Erigeron eriocephalus, Vaccinium uliginosum* ssp. *microphyllum, Carex rupestris, C. nardina, C. supina* ssp. *spaniocarpa, C. norvegica, Calamagrostis purpurascens*

Fig. 44. Stauning Alper, niche N 8, interior of Linné Gletscher (WA2, 1954). [Aerial photo Nr. 4031 of the Danish East Greenland Expeditions 1947-1958, reproduced with permission of the Dansk Geologisk Museum, København, Danmark].

1100 m *Silene acaulis, Chamaenerion latifolium, Luzula spicata*
1030 m *Woodsia glabella, Kobresia myosuroides, Carex capillaris s.l., C. norvegica*

WA2: N 9, Nathorst Land, Ismarken and Højedal, 860-1450 m (Fig. 15)

Nine species reach altitudes of 1450-1500 m near Ismarken. Above 800 m there are a several unexpected collections.

1450 m *Woodsia glabella*
1000 m *Antennaria canescens*
945 m *Hierochloë alpina*
910 m *Potentilla rubricaulis*
900 m *Eriophorum triste, Arctagrostis latifolia, Armeria scabra* ssp. *sibirica* (Fig. 37)
860 m *Dryas octopetala s.l., Draba glabella, Kobresia myosuroides, Carex rupestris, C. nardina, Calamagrostis purpurascens*

WA3; N 10, South Andrée Land: Junctiondal, plateau east of Junctiondal, Månesletten, Gauligletscher, 970-1310 m

The valleys from the regions of South Andrée Land come together in a high plateau to form niche N 10, covering the altitude range from 970 m to 1310 m.

Some calciphytes have been found at high sites on the sedimentary plateau east of Junctiondal with limestone and dolomites.

1310 m *Papaver radicatum s.l., Saxifraga caespitosa s.l., Luzula confusa*
1285 m *Calamagrostis purpurascens, Draba adamsii s.l., D. alpina s.l., D. bellii, Braya purpurascens, Phippsia algida s.l., Poa pratensis s.l.*
1250 m *Saxifraga rivularis s.l.*
1230 m *Taraxacum arcticum,*
1210 m *Colpodium vahlianum*
1200 m *Ranunculus affinis s.l*
1190 m *Cassiope tetragona*
1170 m *Braya linearis*
1150 m *Ranunculus hyperboreus, Potentilla rubricaulis, Pedicularis hirsuta, Eriophorum scheuchzeri, Carex atrofusca, Alopecurus alpinus, Festuca hyperborea, F. vivipara*
1140 m *Poa alpina s.l.*
970 m *Carex maritima, Eriophorum triste*

WA3: N 11, East Andrée Land: Luciadal, 1000-1300 m

Many species reach unusually high altitudes in the Lucia Dal, including

1300 m *Carex supina* ssp. *spaniocarpa*
1230 m *Draba fladnizensis*
1000 m *Antennaria canescens*

WA3: N 12, East Andrée Land, Benjamin Dal, 650-1180 m

Several species have been collected at unusual sites in the middle and high altitude of the interior of Benjamin Dal including

1180 m *Erigeron eriocephalus*
1150 m *Erigeron compositus*
650 m *Minuartia rossii, Pyrola grandiflora, Carex glacialis, C. norvegica, Luzula spicata*

WA3: N 13, North Andrée Land, Moraænedal, Agardh Bjerg 970-1400 m

A few species were collected at high altitudes either along the south-facing slope of Agardh Bjerg or in small niches with favourable conditions along the main glacier and in the mountains south of the valley. These include

1400 m *Campanula uniflora*
1350 m *Potentilla hyparctica*
1250 m *Arenaria pseudofrigida, Hierochloë alpina, Phippsia algida s.l.*
1220 m *Cassiope tetragona, Carex nardina*

1100 m *Ranunculus glacialis* (Fig. 38)
1080 m *Carex bigelowii s.l.,*
1050 m *Arnica angustifolia*
970 m *Ranunculus affinis s.l., Pyrola grandiflora*

WA4: N 14, Waltershausen Nunatak

Among the 20 species collected on the Waltershausen Nunatak are a few species unexpected at this altitude: *Ranunculus glacialis, Potentilla hyparctica, Erigeron eriocephalus, Hierochloë alpina.*

WA4: N 15, Northern side of Bernhard Studers Land

Among the large number of fell-field plants a few basiphilous species have reached the altitude above 1150 m: *Braya purpurascens, Draba bellii, D. arctica s.l., D. nivalis, D. subcapitata, D. glabella, Chamaenerion latifolium, Erigeron eriocephalus, Trisetum spicatum.*

WA4: N 16, North-western Bartholin Land

A few species of the continental föhn-steppe have reached this site in northwestern Bartholin Land: *Potentilla hookeriana s.l., P. hyparctica, Draba fladnizensis, Taraxacum phymatocarpum, Kobresia myosuroides, Carex nardina, C. rupestris, C. misandra, Festuca baffinensis.*

WA5: N 17, Kronprins Christian Land, Drømmebjerg, 600-700 m

More than 40 species were found at 600-700 m at Drømmebjerg due to the ecological variety of the sites and due to the influence of the temperature inversions in this region not far away from the coast.

700 m *Eriophorum scheuchzeri, Minuartia rossii, Phippsia algida s.l.*
670 m *Potentilla hyparctica, Cassiope tetragona, Taraxacum arcticum, Carex rupestris, C. misandra, Festuca brachyhylla, F. hyperborea, Trisetum spicatum*
650 m *Potentilla hookeriana s.l., Draba bellii, D. arctica s.l., Carex nardina, C. glacialis, Poa abbreviata, P. glauca*

WA5: N 18, Ingolf Fjord, 330-770 m

There are a few collections from the northern side of Ingolf Fjord where a considerable number of species have been collected at levels above 300 m due to the effect of the temperature inversions near the fjord.

470 m *Cassiope tetragona, Potentilla rubricaulis, Taraxacum arctogenum, Poa arctica*
410 m *Carex nardina*
330 m *Draba bellii, Calamagrostis purpurascens, Carex rupestris*

11.4 Discussion of questions raised in 11.3.1: B 1, B 2, and D1 -D8

11.4.1 B 1) Influence of latitude and altitude on the number of species

The number of species is reduced with higher northern latitude und high altitude as it is shown in chapter 4, Tab. 9.

11.4.2 B 2) Influence of mountain mass elevation and increasing continentality

It is well known that altitude limits of the tree-line and the altitude limit of vascular plants rise to higher levels in extended mountain massifs (see e. g. Brockmann-Jerosch 1919, Gams 1931, 1932, van Steenis 1961).

The effect of the mountain mass is remarkable in the central and western Stauning Alper (WA2, N 4-8). The influence may also be increased by the effect of continentality. Compared with the southern Werner Bjerge in WA1 the altitude limit of vascular plants is approximately 400 m higher.

A second example – at a smaller scale – is the high plateau of sediments in South Andrée Land (WA3, N 10) with altitudes between 1100 m and 1400 m, where many species reach their altitude limits.

11.4.3 D 1) Inversions of summer temperatures near the coast

At the coast of Kong Oscar Fjord (71°N), as well as along Ingolf Fjord and Hekla Sund (80°N) inversions of temperature are quite common during the summer. Fog or low clouds reduce the temperature and increase the humidity at low altitude, creating unfavourable conditions particularly in the north of the fjord for the development of the vegetation. Therefore, the number of species is smaller in the lowest altitude band than at higher altitudes above the inversion layer.

The effect of temperature inversions was noticed for the first time by the author in the late summer 1951, when working on Syltoppene at the coast of Kong Oscar Fjord.

The influence of temperature inversions is still more conspicuous in WA5. The richest vegetation was found at altitudes above 300 m. A good example is given by the niche N 17 (WA5, Drømmebjerg), where 47 species were collected at altitudes between 600 m and 700 m.

11.4.4 D 2) Effects of local glaciers

Where glaciers are present in an area the temperatures during the growing season are lower than in adjacent areas without glaciers. This decrease of temperature is shown by the lower altitude limits of species in regions where glaciers are present.

Near the ice-cap Ismarken in Nathorst Land (WA2, N 9) the vegetation-line and the altitude limits of the most common species are about 200-300 m lower than in the Stauning Alper in spite of the considerable mountain mass of western Nathorst Land.

The best example to demonstrate the influence of glaciers on the altitude distribution of species has been found in the nunataks north of 74°N. The highest sites are found in Bartholin Land, where a few species reach the altitude of 1470 m (Schwarzenbach 1961), and where some species of the continental föhn-steppe were found at 1180 m (see WA4, N 16). The altitude limits are lower in Bernhard Studer Land, in North Strindberg Land and on Waltershausen Nunatak.

11.4.5 D 3) Influence of special ecological conditions of the habitat

Salinity of soils. Species adapted to a high content of mineral-salts in the soil have been collected in extremely dry niches in the western continental belt which extends east of the inland ice from Scoresby Sund to Ole Rømer Land. These sites are exposed to the föhn-winds desiccating the soil and causing the accumulation of salt on the surface of the soil. To give en example: *Armeria scabra* ssp. *sibirica* was found at an altitude of 900 m near Ismarken in Nathorst Land (see 11.3.2, WA2, N 9, Fig. 37). A second example is the presence of basiphilous species in dry niches of the crystalline parts of the Stauning Alper (see 11.3.1, WA2, N 8 and N 4).

Calciphytes and species of the continental föhn-steppe have also been collected in the crystalline parts of the westernmost nunataks north of 74°N: WA4, N 15, N 16 (Schwarzenbach 1961).

11.4.6 D 4) Taxonomic, cytological or ecological sub-units adapted to special conditions of their habitat

The geographic segregation of different chromosomal races and specifically adapted ecotypes has often been discussed in the literature (e. g. Böcher 1951 b, 1959, 1966, Merxmüller 1952, Favarger 1972, 1975, Holt 1990). A special interest should be given to species with an apomictic reproduction as observed in the genera *Ranunculus, Potentilla, Antennaria* (Porsild 1965), or *Taraxacum*. The following list demonstrate possible genetic or ecological adaptation to high altitude sites.

Antennaria canescens
It is possible based on herbarium specimens that an apomictic race of *A. canescens* (Porsild 1965) has reached a closed area of high altitude in the central and western parts of the Stauning Alper (WA2, N 4, N 6 – N 8).

Antennaria porsildii
A. porsildii was found together with *A. canescens* in WA1 and WA2. As the species has been collected along Sefstrøm Gletscher at the altitude of 1010 m, it is a possible that an apomictic race may occur in the Stauning Alper, which is specially adapted to the severe conditions of high altitude.

Cochlearia groenlandica
Böcher et al. (1978) mention the possibility that a separate race of *C. groenlandica* is found near snow-patches and on solifluction soils in the inland. This hypothesis is supported by three collections from WA1 and from WA5: The species was collected at 510 m near Pingo Pass and at 1000 m in the niche WA2, N 2, south of Randspids. A similar observation has been made by C. Bay (pers. comm.) Furthermore, *C. groenlandica* has been collected at 250 m around Centrum Sø in Kronprins Christian Land (80°N).

Draba alpina s.l.
The species has been found in WA1-WA5. *D. alpina s.l.* has been collected in niche WA1, N 1, south of Randspids: 1200 m and in niche WA3, N 10, Gauligletscher: 1285 m, Månesletten: 1180 m. It is possible that these specimens belong to an ecotype adapted to high altitude.

Draba nivalis
There are 13 specimens from WA1 (southern Werner Bjerge) and from WA2 (Stauning Alper) at high altitudes between 930 m and 1720 m marking an apparently closed area of *D. nivalis*.

Festuca vivipara
Two specimens of *F. vivipara* have been found at an unexpected altitude in niche WA3, N 10, WA3, east of Junctiondal: 1150 m and Junctiondal: 1100 m.

Luzula spicata
Luzula spicata has been found in an area of high altitude in WA2, N 4, N 7 and N 8, central and western Stauning Alper, as well as in WA3, N 12, Benjamin Dal.

Phippsia algida ssp. *algida. and P. algida* ssp. *algidiformis*
Phippsia algida ssp. *algida* was found in all five study areas, following in WA3-WA5 the altitude pattern of ADT C1. The species was rather rare and has been collected only at lower altitudes (WA2, near Ismarken: 800 m; WA1, around Pingo Pass: 620 m). Two specimens from WA1, Schuchert Dal: 300 m, Pingo Pass: 410 m have been determined as ssp. *algidiformis* (Bay 1993).

Poa alpina s.l.
Schwarzenbach (1961) has described a variety *P. alpina* var. *saxicola* distinguished morphologically from *P. alpina* var. *typica.* The seminiferous *P. alpina* var. *typica* has been found in herb-slopes of WA1-WA4 at low altitude (highest site: WA1, N 3, Biskop Alfs Dal: 770 m). The variety *P. alpina* var. *saxicola* was collected at high altitudes in three study areas (WA2, N 5, Vikingebræ: 810 m, N 7, east of Skjoldungebræ: 1330 m, Sedgwick Gletscher: 780 m; WA3, N 10, Gauligletscher: 1140 m; WA4, Vibeke Nunatak: 920 m).

Poa pratensis s.l.
The collective species *P. pratensis s.l.* (including the viviparous var. *colpodea*) is composed of several small taxa differentiated by the numbers of chromosomes. The non-viviparous *P. pratensis s.l.* was found at sites of low altitudes in WA1-WA5 with the exception of one site in WA3 (N 10, Månesletten: 1285 m). This specimen may belong to an ecological or cytological race.

Potentilla hyparctica
Seven collections of *P. hyparctica* are from WA4. As the nunataks north of 74°N belongs to an important niche, all sites are listed (WA4, Waltershausen Nunatak: 1270 m, Krumme Langsø: 200 m, Bartholin Land: 880 m, 900 m, 1180 m, Vibeke Nunatak: 900 m Vibeke Sø: 780 m).

Potentilla rubricaulis
A well consolidated population of *P. rubricaulis* has been found in WA3, North Andrée Land, Morænedal, where seven specimens have been collected between 160 m and 850 m. The highest site is south of the main glacier.

Saxifraga nathorstii
S. nathorstii is a polyploid hybrid between *S. oppositifolia* and *S. aizoides* endemic in East Greenland. The species has been collected in WA1, WA3, and WA4. It is in general a species of low altitude with a similar altitude distribution as *S. aizoides.* However, there are a few collections from altitudes above 400 m in an area between WA4, Morænedal and the nunataks north of 74°N (WA4, Vibeke Sø: 780 m, Bartholin Nunatak: 900 m, C. H. Ostenfeld Nunatak: 415 m, 640 m, 700 m).

Taraxacum arcticum
This species was collected in all five study areas. *T. arcticum* has been found generally below 1000 m with two exceptions (WA2, N 4, Sefstrøm Gletscher: 1230 m; WA3, N 10, Junctiondal: 1230 m). As it is well known many small-species occur within the genus *Taraxacum* due to the apomictic method of reproduction, it is possible therefore that such an apomictic race of *T. arcticum* occurs in East Greenland which is well adapted to the severe ecological conditions of high altitudes.

11.4.7 D 5) Niches as pioneer sites of immigrating and migrating species

While the hypothesis about the survival of species which immigrated before the latest big advance of the ice has often been discussed, the question of immigration in recent times has only been considered occasionally. An interesting example has been given by Halliday et al. (1974). Here they discuss the isolated occurrence of some southern species near the hot springs of Blosseville Kyst. They suggest that propagules of these species have been transported form the south by geese stopping at this place during their flight to the breeding areas further north.

Observations on the revegetation of an old airstrip and dirt-roads near Pingo Pass (WA1) have shown that about 2/3 of all species found on the plateau of the Pingo Pass had successfully recolonized the open grounds of the airstrip and the dirt roads within the period 1957/1991 (Schwarzenbach 1996 a). This result shows that the migration of plants in the Arctic can take place relatively quickly.

Propagules of plants might be transported by wind, by drifting and melting snow, and by streams as well as by birds and mammals over short and long distances. Seeds of *Salix arctica* or *Chamaenerion latifolium,* and of the genera *Dryas, Eriophorum, Erigeron and Taraxacum* have often been observed by the author when they drifted with strong winds over summits and ridges at altitudes of more then 1500 m.

The most striking example of migration was the observation of just one individual of *Saxifraga cernua* during the first ascent of Elisabeth Bjerg (Figs. 8, 11) in the Central Stauning Alper. The plant was growing at 2200 m in a rock-fissure of the long ridge leading to the summit and was in flower.

It is well known that some species are often found on ,manured' ground around bird perches (Seidenfaden & Sørensen 1937, Bay 1992). It is assumed that propagules of these species could be transported by birds to places at high altitude. However, in spite of the fact that many bird perches have been visited, no pioneer plants have been found around bird perches at high altitude.

On the other hand there are a few examples supporting the assumption that propagules of some species might have been carried by birds to remote sites. In WA1, N2, north of Pingo Pass the author observed flocks of geese visiting a place at an altitude of 1000 m, where the snow melted late. *Cochlearia groenlandica, Equisetum variegatum, Saxifraga aizoides* and *Luzula arctica* have been collected there. The plants on the wet creeping soil were in full development at the end of July. In WA1, south of Randspids: 1200 m the species D. *alpina s.l., D. bellii, D. gredinii, Polygonum viviparum, Saxifraga cernua, Carex maritima, Trisetum spicatum* have been found. As Gelting (1937) has stated that the East Greenland ptarmigan often feed on the bulbils of *P. viviparum* and *S. cernua,* it is quite probable that propagules of these two species have been transported by ptarmigan.

11.4.8 D 6) Refugial niches of southern species

Several observations support the hypothesis that populations of southern species, which have reached high altitudes during a period with a warmer climate, might have survived in suitable niches, as it is shown by the following list. There is one question to be kept in mind: It should be unlikely that the niche has been colonised recently from below.

WA1, N 3, west of Biskop Alf Dal: 770 m

A herb-slope with many species has been observed at an altitude of 770 m (*Botrychium lunaria, Thalictrum alpinum, Sibbaldia procumbens, Gentiana tenella, Antennaria canescens, Arnica angustifolia, Taraxacum brachyceras, Luzula spicata).*

WA2, N 4, Sefstrøm Gletscher: 960-1260 m

There are a few surprising collections of southern species along Sefstrøm Gletscher. Some of the species belong to a group of continental species as found in Hjørnedal, Gåseland, 70°N in the summer 1994 (Schwarzenbach, unpublished): *Sibbaldia procumbens* (three sites between 1010 m and 1030 m), *Euphrasia frigida* (four sites between 1010 m and 1230 m, *Harrimanella hypnoides* (1030 m), *Taraxacum brachyceras* (1030 m), *Luzula spicata* (nine sites between 960 m and 1260 m)

In similar niches a few species of herb-slopes and heaths have been found at high sites: *Antennaria canescens* (three sites between 960 m and 1200 m), *A. porsildii* (1010 m), *Arnica angustifolia* (960 m and 1230 m), *Campanula gieseckiana s.l.* (1200 m), *C. uniflora* (1200 m), *Carex scirpoidea* (960 m), *Empetrum nigrum* var. *hermaphroditum* (960 m), *Erigeron humilis* (960 m, 1010 m, 1030 m), *Hierochloë alpina* (1230 m), *Pedicularis flammea* (1230 m), *Tofieldia pusilla* (960 m, 1030 m).

WA2, N 5, northern side of Vikingebræ: 360-870 m (Fig. 43)

Several southern species have been collected along the slope north of Vikingebræ: *Betula nana* (870 m), *Carex scirpoidea* (870 m), *C. bigelowii* s.l. (850 m), *Tofieldia pusilla* (840 m), *Potentilla crantzii* (810 m), *Sibbaldia procumbens* (810 m), *Viscaria alpina* (150 m, 810 m, Fig. 36), *Euphrasia frigida* (810 m), *Antennaria canescens* (810 m), *Taraxacum brachyceras* (360 m, 810 m), *Luzula spicata* (810 m), *Arabis holboellii* (360 m, Fig. 33), *Draba aurea* (150 m, 360 m, Fig. 34).

WA2, N 7, Skjoldungebræ

Luzula spicata (1660 m), *Campanula gieseckiana s.l.* (1320 m), *Antennaria canescens* (1320 m), *Carex atrofusca* (1310 m), *Huperzia selago* (1240 m).

WA2, N 8, Linné Gletscher (Fig. 44)

Campanula gieseckiana s.l. (1335 m), *Antennaria canescens* (1335 m), *Luzula spicata* (1335 m), *Draba glabella* (1150 m), *Vaccinium uliginosum* ssp. *microphyllum* (1120 m), *Carex supina* ssp. *spaniocarpa* (1120 m), *Carex capillaris s.l.* (1030 m), *C. norvegica* (920 , 1030 m, 1120 m).

11.4.9 D 7) Refugial niches of northern species

If northern species have migrated to East Greenland during advances of the ice and have occupied a more or less closed area at low or middle altitude, they have probably moved to higher levels during later periods when temperatures were higher. As a result of these climatic changes many northern species might now be found in niches at high altitude. The following niches may belong to this kind of possible refugia: WA1, N1, south of Randspids: 1200-1370 m; WA2, N 4, Sefstrøm Gletscher: 1300-1670 m; WA2, N 6, Gullygletscher: 1460-1670 m; WA2, N 7, Skjoldungebræ: 1700-1720 m; WA3, N 10, Junctiondal, plateau east of Junctiondal, Månesletten, Gauligletscher: 1150-1285 m; WA3, N 13, Moræenedal, Agardh Bjerg: 1100-1500 m; WA4, N 16, Bartholin Land: 1350-1470 m.

11.4.10 D 8) Disjunct patterns of altitude and geographic distribution

There are several species with a disjunct altitude and geographic distribution within the study areas These patterns are described, but there is no attempt to explain them.

Arenaria pseudofrigida

A. pseudofrigida was collected in WA1, WA3 and WA4. The species follows regions of the continental belt. Two sites are above 1000 m: WA1, N 1, mountains south of Randspids: 1100 m; WA3, N 13, Agardh Bjerg: 1250 m. There are two more sites in WA4 with altitudes of 350 m and 450 m.

Braya linearis
Due to the ecological demand the rare species *B. linearis* has only been collected in the western continental belt in WA3 and WA4. There is one site at the unusual altitude of 1170 m in South Andrée Land on the plateau east of Junctiondal (N 10).

Carex glacialis
There are three small regions at rather high altitude, where *C. glacialis* has been collected: WA2, N 8, Linné Gletscher (1030 m, 1300 m; WA3, N 12, Benjamin Dal (560 m, 650 m); WA5, N 17, Drømmebjerg (670 m).

Carex norvegica
C. norvegica has been found in three small regions only. These three sites in WA1-WA3 are probably outposts of this southern species: WA1, Kong Oscar Fjord, Syltoppene (210 m); WA2, N 8, Linné Gletscher (920 m, 1030 m, 1120 m); WA3, N 12, Benjamin Dal (650 m).

Draba arctogena
Three specimens have been collected at levels between 900 m and 1370 m in WA4, one plant was found in WA3, N 10, Gauligletscher: 1150 m. These collections suggest, that the northern *D. arctogena* is distributed in the western continental belt north of 73°N. Further observations are needed to clarify the geographic and altitude distribution of this species in Northeast Greenland.

Erigeron compositus
E. compositus was found at two sites only, occurring in isolated niches at rather high altitudes: WA2, N 8, Linné Gletscher: 1150 m; WA3, N 12, Benjamin Dal: 1150 m.

Huperzia selago
H. selago is often found in the heath dominated by *Cassiope tetragona* and has been collected in WA1-WA3 at low and middle altitude. However, there are two exceptions: WA2 2, N 4, Sefstrøm Gletscher: 1030 m, N 7, Skjoldungebræ: 1240 m.

Pyrola grandiflora
The gregarious species *P. grandiflora* shows a somewhat scattered occurrence in WA1-WA4. The species was quite common in a few regions such as WA2, Ismarken and WA3, Benjamin Dal. *P. grandiflora* was collected only at alti-

tudes below 1000 m, but has once been observed in WA2, N 9, Ismarken: 1400 m.

Ranunculus glacialis
It seems that *R. glacialis* occupies a more or less isolated area following a belt west of 23°30'W from North Andrée Land (73°46'N) to Vibeke Nunatak at 74°27'N with an altitude range between 780 m and 1350 m.

R. glacialis prefers crystalline rocks and it often grows on wet soil in and below snow-patches (Schwarzenbach 1961). The species has been collected in WA3 and WA4. It has been found at rather high altitudes (WA3, N 13, Agardh Bjerg: 1100 m; WA4, N 14, Waltershausen Nunatak: 1270 m, North Strindberg Land: 960 m, western Bartholin Land: 1350 m, Bartholin Nunatak: 900 m, Vibeke Nunatak : 900 m).

Bay (1992) shows that *R. glacialis* has been found also in the westernmost belt further to the north. The author does not know how this peculiar distribution of the species in the mountains near the Inland Ice might be explained.

Ranunculus pygmaeus
R. pygmaeus was found generally at low altitude in WA1-WA3. There is one remarkable exception: WA2, N 4, Sefstrøm Gletscher: 1200 m.

Salix herbacea
S. herbacea has occasionally been collected at low altitude in WA2 and WA3. However, there is one collection from a high niche: WA2, N 4, Sefstrøm Gletscher: 1030 m.

Saxifraga rivularis s.l.
S. rivularis s.l. was collected in WA1-WA4 at low and middle altitude with one exception: WA3, N 10, east of Junctiondal: 1250 m.

11.5 Discussion

11.5.1 Analysis of the altitude distribution of vascular plants

Several possibilities must be kept in mind if the influence of climate and climatic changes on the arctic mountain vegetation is studied. The analysis of the geographic and altitude distribution of individual species has brought some interesting and unexpected findings:

- Three examples show that there is a considerable influence of the mountain mass on the altitude distribution of vascular plants in the mountains of East and Northeast Greenland (11.4.2).

- The altitude limit of vascular plants becomes lower (11.4.4) near local ice-caps, near the Inland Ice and around ice-streams with many tributary side-glaciers.
- The salinity of soils depending on the effects of drying winds in summer, results in the immigration of halophytes and species of the föhn-steppe into the western continental belt between the inner fjords and the Inland Ice. The high evaporation with the resulting accumulation of mineral-salts at the surface of soils enables basiphilous species to reach isolated sites in the crystalline parts of the Stauning Alper and of the nunataks north of 74°N (11.4.5).
- Inversions of temperatures during the growing season have a remarkable effect on the altitude distribution of species near the outer coast north of 80°N and, to a less degree along the shore of Kong Oscar Fjord (11.4.3).
- Some examples support the hypothesis that taxonomic or cytological races as well as ecotypes occur in some regions or have reached niches at high altitudes. Genera with apomictic types of reproduction, such as *Antennaria, Taraxacum* or *Potentilla* contribute to the creation of isolated populations (11.4.6).

If a species is found in a niche of high altitude there are three possibilities:

- Are these niches pioneer-sites at high altitude reached by the species during its recent immigration in the area?
- Are these niches refugia, which have been reached by a population of the (southern) species during a period with a warmer climate and where the species might have survived the following advances of the ice?
- Are these niches refugia, which have been reached during a period with a warmer climate by a population of the (northern or circumpolar) species immigrating during an earlier advance of the ice?

11.5.2 Pioneer-sites

As the examples in chapter 11.4.7 show, there is no doubt that some species have reached by chance isolated sites at an unusually high altitude, for example if their propagules have been transported from lower altitudes by wind, snow or birds. If the local conditions are unfavourable the species will not survive and will disappear within a short time. If the new site offers a suitable habitat the species will establish a population able to survive as a pioneer species at high altitude.

In discussing the question of a possible survival of southern species which could have migrated before the latest advance of ice the possibility of a recent migration has often been overlooked.

11.5.3 Refugial sites of southern species

Species immigrating in periods with a warmer climate can survive in niches if they are able to maintain self-sustaining populations. Therefore, it is necessary to check in the field if these conditions are fulfilled. The required number of individuals for the sustained regeneration of the population depends on the life-cycle and the ecological strategy of the individual species. To give an example: The *Gentiana tenella* seems to be able to maintain its population by colonies of a few hundred individuals only.

There are a number of examples supporting the hypothesis that species which have migrated in times with a warmer climate have survived till now in niches at surprisingly high altitudes. The highest sites have been found in the central and western Stauning Alper and in the more or less glacier free regions of East Andrée Land. The niches near the edge of the Inland Ice do not reach high altitudes due to the decrease in temperature near the local ice-caps and in the interior parts of the big glaciers flowing eastwards from the Inland Ice. A further refugium could be the belt of mountains and nunataks between ca. 24°00'W and the Inland Ice from 73°30'N to 74°30'N.

No new information has been gained about the time of the presumed migration of species surviving at high altitudes. There are no new arguments to suggest whether or not these species migrated from south to north during an interglacial period or during a warmer postglacial period.

11.5.4 Refugial sites of northern species

The question has been asked whether species immigrating from the north during a colder period could have survived in certain niches. It is obvious that species best adapted to a cold climate can grow under the severe conditions near the altitude limit of vascular plants. If temperatures increase during a long period this limit rises to higher levels and the species of fell-field and snow-patches reach sites at higher altitudes. At the same time grasses, sedges and dwarf-shrubs move upwards competing with established plants typical of the band near the vegetation limit. However, there are always niches with unfavourable conditions at low altitudes where the species of high altitude can grow.

The conclusion: If species of higher altitude are found at low altitudes, it is not clear whether these species have migrated during a colder period or have just reached and remained in sites with unfavourable conditions where species demanding better conditions cannot compete.

On the other hand it is to be expected that northern and circumgreenlandic species are found in the highest niches of all five study-areas. The occurrence of high arctic snow-patch plants at high levels in southern regions is explained by the fact that accumulations of snow near the firn-line melt late.

11.5.5 Possible influence of climatic changes on the vegetation based on the analysis of the present altitude distribution of plants and other factors

It is possible to predict possible climatic changes in Greenland vegetation using an analysis of altitude distribution in areas with different conditions of temperature and precipitation. The question, how is the present altitude distribution of species effected by macro and micro climate, is key to the prediction of the influence of climate change.

It is evident that the living conditions of the micro habitat have a decisive influence on the development of a plants, constraining the effects of macro-conditions. The temperature of the soil and of the air might vary considerably just above the ground within distances of a few centimetres. The supply of water depends more on the accumulation of snow during the winter than on the rain during the growing season. The deposition of fine material washed down from an area with a different chemical composition may reduce the influence of the local bedrock. These are all important micro conditions.

However, in spite of these factors, the influence of the microclimate is limited. Analysing the pattern of altitude distribution based on a statistical approach it is quite possible to show the effects of macroclimatic factors. In chapter 8 statistical methods of estimating the altitude limit of vascular plants are discussed.

Influence of presumed changes of temperature

The observed differences between the altitude limits of the individual species in different regions have been explained by the difference of temperature during the growing season (see also Sørensen 1941). The altitude limits of the most common species show a considerable variation from one region to the other. If the five study areas are compared, the difference between the average altitude of the most common species (Tab. 45) has a range of 874 m (comparison between WA2/WA5). Assuming a temperature gradient of 0.5°C/100 m this difference of 874 m is equivalent to a difference of more than 4°C during the growing season.

Based on the analysis of the altitude distribution a possible rise of the mean temperature by about 4°C could influence the pattern of the present vegetation in East and Northeast Greenland as follows:

- The altitude limit of vascular plants, and the altitude limits of the individual species would be several hundred meters higher than today.

- The relative sequence of the specific altitude limits would not change.
- The same types of vegetation would be found as nowadays, but the zone of snow-patches and constantly irrigated fields below snow-beds would be found at higher levels.

It might be possible that species from further south in Greenland could migrate and occupy the lowest altitude bands. In the area between 71°N and 73°N it would be probable that a number of southern species now found in Jameson Land would reach new areas further north. A possible migration route could follow the sedimentary belt east of the crystalline zone of the Werner Bjerge and the Stauning Alper. A second route of immigration might be used by species from the continental part of the interior Scoresby Sund following the passes between Schuchert Dal and Alpefjord or along the passes leading via Furesø. As pointed out in chapter 11.4.8 there are observations of isolated southern species found along these supposed routes of migration.

Influence of possible changes of precipitation.

If a considerable increase of the annual amount of precipitation occurred, the effect on the vegetation would depend on the resulting pattern of snow-accumulation. If the number of snow-patches and snow-beds as well as the accumulated volume of snow increased the vegetation of the high altitude bands would probably have a similar character to that of the present vegetation at the outer coast. A reduction of precipitation would probably lead to an extent of the dry areas in the interior and middle parts of the East Greenland fjords.

An increase of the annual precipitation could also change the present zonation of continental and oceanic belts. It is not assumed that vegetation-types such as the continental föhn-steppe would disappear or become restricted to isolated niches.

Influence of possible changes of the wind-system.

The present system of seasonal winds influences the vegetation much more than has been assumed hitherto. Several aspects have to be kept in mind:

- Snow-drifts depend on the force and the direction of winds prevailing in winter-time. The local pattern of snow-patches and snow-beds in spring-time demonstrates wind conditions and precipitation from the preceding winter. The development of plants depends to a high degree on the amount of melt-water available during the growing period. Therefore, the varied distribution of snow by the wind is an important ecological factor influencing the local growth of the vegetation.
- The dry and warm föhn-winds blowing from the Inland Ice in a western direction cause early melting of the snow-cover and lead to an intense desiccation of the soils during summer. They initiate a process of deser-

tification including the accumulation of mineral-salts at the surface of the soil.

- The foehn-winds prevent the cool and humid air above the Polar Sea reaching the middle and interior parts of the East Greenland fjord-systems.
- As it is difficult to imagine how the present wind-systems might be influenced by the assumed change of climate it is difficult to speculate on the possible changes of vegetation in East and Northeast Greenland mountains.

General remark

In the mountains of East and Northeast Greenland the distribution of species depends to a great extent on the local conditions of the micro-habitat changing often within very small distances. Observations in the field have shown that changes in macroclimatic conditions – as they are postulated for the near future – might have only a slight effect on arctic mountain vegetation. The numbers of species and their local distribution would not change very much. This study of Greenland confirms the conclusions of botanists studying the influence of possible climatic changes on the vegetation of the Alps (Moore 1990, Tallis 1991, Grabherr et al. 1994, 1995, Gottfried et al. 1994, Körner 1994, Theurillat 1995).

12 Summary

This study sets out the author's observations of vascular plants in mountainous areas of East and Northeast Greenland between 71°30'N and 80°30'N during the summers 1948-1952, 1954, 1956 and 1991. Some results of three other expeditions to Gauss Halvø, 1948, 73°N, to Gåseland, 1994, 70°N and to North Peary Land, 1995, 83°N have been used for comparison.

The field work in many regions of East and Northeast Greenland between the outer coast and the nunataks near the Inland Ice provided an opportunity to study the altitude distribution of vascular plants of the High Arctic, where the vegetation is aneffected by human activities. Therefore, the present altitude and geographic distribution of plants allows to study the influence of the natural conditions of environment. The altitude distribution is described for all 176 species of vascular plants found in five study areas (WA1-WA5). Of these species, 107 have altitude limits at or above 1000 m, and 50% of all species reach at least 1150 m. The highest record *(Saxifraga cernua)* is from 2200 m (Elisabeth Tinde, Stauning Alper, 72°N).

There is a considerable rise in altitude limit of vascular plants from the outer coast in the East to the innermost mountain chains in the West. In the Stauning Alper (72° N) the altitude limit of vascular plants is about 300 m higher in the west compared with the mountains along the south-western coast of Kong Oscar Fjord. This difference of the altitude limit corresponds to a temperature difference of c. 1.5° C within a distance of c. 50 km.

The altitude limit (based on herbarium specimens only) decreases considerably from the highest level in the Stauning Alper (WA2 ,72°N, 1720 m) to Kronprins Christian Land (WA5, 80°N, 900 m). The difference of 820 m corresponds to a decrease of c. 4.2° C. There is also a considerable influence from exposure, as more species are found on sites facing SE-SW exposures and fewer facing NE-NW. The altitude average for the first group (based on 20 species) is 1370 m, and 799 m for the second group. This difference corresponds to a difference in summer temperature of c. 2.4° C.

The pattern of altitude and latitude distribution as well as the habitat of each species is used to develop a classification of ,Altitude Distribution Types (ADT)'. This system is compared with the ,North Greenland Distribution Types (NGDT)'as defined by Bay (1992). Species found in the highest altitude belt belong mainly to the species with a North Greenland distribution. Species with their northern limit south of 76°N reach their altitude limit in the lowest altitude belts of WA1-WA4.

The altitude distribution of individual species within the five study areas has been studied in detail based on statistical methods. A fairly large number of species occur in isolated sites at high altitude. It is assumed that some of these sites serve as refugia for species better adaqpted to climate elsewhere. Methods for the estimation of altitude limits have been developed and compared. Several of these methods have been combined in order to compare altitude limits in different areas. A useful method for comparison is based on the observed altitude limits of 42 common species. This has been used for comparison between the five study areas and also with six areas in other parts of West, East and North Greenland.

Species diversity based on the "species/presence-distribution" show a rather homogenous vegetation in the whole study area (WA1-WA5).

The author hopes that this study will stimulate further studies and discussion on the altitude distribution of vascular plants in other parts of Greenland and on mountains outside the Arctic. It would be of special interest to be able to relate changes in altitude distribution with climate changes due to the "Greenhouse effect".

13 Acknowledgements

The study could be only realised in international and interdisciplinary co-operation with many partners. As the project began in 1948 some of these colleagues died in the mean-time. Nevertheless, the author wishes to include them in the acknowledgements as they have contributed to the realisation of this long-lasting work.

Dr Lauge Koch enabled the author to participate as a member of the Danish East Greenland Expeditions 1948-1952, 1954 and 1956 organised by an experienced and efficient staff. The writer thanks the colleagues who were partners in the field work of the eight unforgettable summers: H. Bütler, B. Hinsch (1948), E. Fränkl (1949-1952), P. Braun (1950, 1951), F. Müller (1952), J. Cowie (1952), P. Adams (1952), J. Haller (1954, 1956), W. Diehl (1954). G. Jensen, Expeditions Adviser, Farum (DK) has competently organised the expedition of 1991 to the South Werner Bjerge. R. Burton, Arcturus Expeditions, Gartochan (U.K.) lead the expeditions to Gåse Land (1994) and to North Peary Land (1995). My wife Elisabeth Schwarzenbach has joined me during the expeditions of 1991, 1994 and 1995.

The botanists of the Botanical Museum and of the University of Copenhagen familiar with the many-sided problems of Greenland botany identified the author's collections. Their advice based on experience from their own fieldwork in Greenland contributed a lot to the present study. The writer thanks J. Grøntved, O. Hagerup, Th. Sørensen, K. Holmen , B. Fredskild and C. Bay for their help. He also includes G. Halliday (Lancaster University) who has identified the author's collections from 1951 and 1956.

The author is grateful to W. Urfer, S. Kaiser and R. Geißdörfer (Department of Statistics, University of Dortmund, Germany) for their valuable co-operation. They treated the statistical problems connected with the estimation of the altitude limit of vascular plants and with the analysis of diversity of the vegetation in Greenland mountains.

B. Fredskild and C. Bay commented the drafts of the present paper and made many valuable proposals. Dr Jean Balfour, Scotland, U.K. revised the language and suggested some additional points based on her own botanical experience from the Arctic. B. La Roche prepared the maps and illustrations.

G. S. Mogensen reviewed the manuscript in detail. His valuable proposals helped to clarify many aspects and to improve the text. I thank him very much for his time-consuming work and his great help.

I thank Kirsten Caning for her competent, co-operative, and efficient help as Publishing Editor.

14. References

Anonymous. 1989. Nuna Tek. 1989. Klima. Jameson Land. Del I, Tekst og tabeller.- Sektionen for hydrotekniske Underdøgelser, København: 17 pp.

Adams, P. J. & Cowie, J. W. 1953. A geological reconnaissance of the region around the inner parts of Danmarks Fjord, Northeast Greenland. Meddr Grønland 111 (7): 24 pp.

Bay, C. 1983. En plantegeografisk undersøgelse af Nordvestgrønland. – Thesis, University of Copenhagen: 238 pp.

Bay, C. 1992. A phytogeographical study of the vascular plants of northern Greenland – north of 74° northern latitude. – Meddr Grønland, Biosci. 36: 102 pp.

Bay, C. 1993. Taxa of vascular plants new to the flora of Greenland. – Nordic Journal of Botany 13: 247-252.

Bay, C. & Boertmann, D. 1989. Biologisk-arkaeologisk kortlægning af Grønlands Østkyst mellem 75°N og 79°30'N. Del 1: Flyrekognoscering mellem Mesters Vig (72°12'N) og Nordmarken (78°N).- Grønlands Hjemmestyre, Miljø- og Naturforvaltning. Teknisk rapport 4: 63 pp.

Bay, C. & Fredskild, B. 1990 a. Biologisk-arkæologisk kortlægning af Grønlands Østkyst mellem 75°N og 79°30'N. Del 3: Botaniske undersøgelser i området mellem Fligely Fjord (74°50'N) og Nordmarken (77°30'N). – Grønlands Hjemmestyre, Miljø- og Naturforvaltning. Teknisk Rapport 11: 56 pp.

Bay, C. & Fredskild, B. 1990 b. Biologisk-arkæologisk kortlægning af Grønland Østkyst mellem 75°N og 79°30'N. Del 6: Botaniske undersøgelser i området mellem Germania Land (77°N) og Lambert Land (79°10'N). – Grønlands Hjemmestyre, Miljø- og Naturforvaltning. Teknisk Rapport 24: 55 pp.

Bay, C. & Fredskild, B. 1994. Grønlands Botaniske Undersøgelse 1993. – Botanisk Museum, København: 14 pp.

Bay, C. & Holt, S. 1986. Vegetationskortlægning af Jameson Land 1982-86. – Grønlands Fiskeri- og Miljøundersøgelser, København: 40 pp.

Bearth, B. 1959. On the Alkali Massif of the Werner Bjerge in East Greenland. – Meddr Grønland 154 (9): 63 pp.

Bengård, H.-J & Henriksen N. 1991. Geological map of North Greenland 1:1.000.000. – The Geological Survey of Greenland, Copenhagen.

Bierther, W. 1941. Vorläufige Mitteilung über die Geologie des östlichen Scoresbylandes in Nordostgrönland. – Meddr Grønland 114 (6): 20 pp.

Binz, A. & Becherer, A. 1976. Schul- und Exkursionsflora der Schweiz mit Berücksichtigung der Grenzgebiete (16. Aufl.). – Schwabe & Co., Basel: 424 pp.

Böcher, T. W. 1933. Phytogeographical studies of the Greenland flora based on investigations of the coast between Scoresby Sound and Angmagssalik. – Meddr Grønland 104 (3): 56 pp.

Böcher, T. W. 1938. Biological distributional types in the flora of Greenland. A study on the flora and plant-geography of South Greenland and East Greenland between Cape Farewell and Scoresby Sound. – Meddr Grønland 106 (2): 339 pp.

Böcher, T. W. 1951 a. Studies on the distribution of the units within the collective species of *Stellaria longipes*. – Bot. Tidsskr. 48 (4): 401-420.

Böcher, T. W. 1951 b. Distributions of plants in the circumpolar area in relation to ecology and historical factors. – Journal of ecology 39 (2): 376-395.

Böcher, T. W. 1956. Area-limits and isolations of plants in relation to the physiography of the southern parts of Greenland. – Meddr Grønland 124 (1): 40 pp.

Böcher, T. W. 1959. Floristic and ecological studies in Middle West Greenland. – Meddr Grønland 156 (5): 68 pp.

Böcher, T. W. 1963. Phytogeography of middle West Greenland. – Meddr Grøn-land 148 (3): 289 pp.

Böcher, T. W. 1966. Experimental and cytological studies on plant species. IX. Some arctic and montane *Crucifers*. – Biol. Skr. Vid. Selsk. 14 (7): 74 pp.

Böcher, T. W., Fredskild, B., Holmen, K. & Jakobsen, K. 1966. The flora of Greenland (2nd edition). – København: 312 pp.

Böcher, T. W., Fredskild, B. Holmen, K. & Jakobsen, K. 1978. Grønlands Flora (3rd edition). – København: 327 pp.

Brockmann-Jerosch, H. 1919. Baumgrenze und Klimacharakter. Beiträge zur Geobotanischen Landesaufnahme der Schweiz 6: 255 pp.

Buchenau, F. & Focke, W. 1874. Gefässpflanzen. – In: Die zweite Deutsche Nordpolfahrt in den Jahren 1869-70. Vol.2: Wissenschaftliche Ergebnisse. Leipzig.

Callomon, J. H., Donovan D. T. & Trümpy, R. 1972. An annotated map of the Permian and Mesozoic formations of East Greenland. – Meddr Grønland 168 (2): 35 pp.

Defretin-Lefranc, S., Grasmück, K. & Trümpy, R. 1969. Notes on Triassic Stratigraphy and Palaeontology of Northeastern Jameson Land (East Greenland). – Meddr Grønland 168 (2): 121 pp.

Dusén, P. 1901. Zur Kenntnis der Gefässpflanzen Ostgrönlands. – Bih. K. Vet.-Akad. Handl., Bd. 27, Afd. 3, No. 3. Stockholm.

Eberlin, P. 1888. Blomsterplantene i Dansk Østgrønland. – Arch. f. Mat. og Naturv. 12 (Kristiania)

Elkington, T. T. 1965. Studies on the variation of the genus *Dryas* in Greenland. – Meddr Grønland 178 (1): 56 pp.

Fantin, M. 1969. Montagne di Groenlandia: Scoperta, esplorazione, spedizioni d'alpinismo fino ai giorni nostri. – Mario Fantin (ed.). Tamari Editori, Bologna. 367 pp.

Favarger, C. 1972. Endemism in the montane floras of Europe. – In: Valentine D. H. (ed.). Taxonomy, phytogeography and evolution. Academic Press, London: 191-204.

Favarger, C. 1975. Cytotaxonomie et histoire de la flore orophile des Alpes et de quelques autres massifs montagneux d'Europe. – Lejeunia, n.s. 77: 1-45.

Feilberg, J. 1984. A phytogeographical study of South Greenland. Vascular plants. – Meddr Grønland, Biosci. 15: 70 pp.

Fränkl, E. 1951. Die untere Eleonore Bay Formation im Alpefjord. – Meddr Grønland 151 (6): 15 pp.

Fränkl, E. 1953 a. Die geologische Karte von Nord-Scoresby Land (NE-Grönland). – Meddr Grønland 113 (6): 56 pp.

Fränkl, E. 1953 b. Geologische Untersuchungen in Ost-Andrées Land (NE-Grönland). – Meddr Grønland 113 (4): 160 pp.

Fränkl, E. 1954. Vorläufige Mitteilung über die Geologie von Kronprins Christians Land (NE-Grönland, zwischen 80°-81°N und 19°-23°W). – Meddr Grønland 116 (2): 85 pp.

Fränkl, E. 1955. Weitere Beiträge zur Geologie von Kronprins Christians Land (NE-Grönland zwischen 80° und 81°30'N.). – Meddr Grønland 103 (7): 35 pp.

Fredskild, B. 1966. Contributions to the flora of Peary Land, North Greenland. – Meddr Grønland 178 (2): 22 pp.

Fredskild, B. 1973. Studies on the vegetational history of Greenland. – Meddr Grønland 198 (4): 247 pp.

Fredskild, B. 1991. Warming og Grønlands planteverden – fra floristik til vegetationsstudier. – Urt 15 (5): 32-39.

Fredskild, B. 1996 a. Grønlands Botaniske Undersøgelser 1995-1996. – Botanisk Museum København: 21 pp.

Fredskild, B. 1996 b. A phytogeographical study of West Greenland between 62°20'N and 74°00'N. – Meddr Grønland, Biosci. 45, 155 pp.

Fredskild, B. 1996 c. Holocene climatic changes in Greenland. Bjarne Grønnow (ed.). The Paleo-Eskimo cultures of Greenland: 244-151. Dansk Polar Center.

Fredskild, B. & Bay, C. 1989. Grønlands Botaniske Undersøgelse 1988. – Botanisk Museum, København.

Fredskild, B. & Bay, C. 1990. Grønlands Botaniske Undersøgelse 1989. – Botanisk Museum, København: 30 pp.

Fredskild, B., Bay, C. & Holt, S. 1982. Botaniske Undersøgelse på Jameson Land 1982. – Botanisk Museum, København: 37 pp.

Fredskild, B., Bay, C., Holt, S. & Nielsen, B. 1986. Grønlands Botaniske Undersøgelse 1985. – Botanisk Museum, København: 40 pp.

Funder, S. 1989 a. Le Quaternaire du Groenland oriental. Chapitre 13. Le Quaternaire des régions extraglaciaires du Groenland et des plates-formes continentales adjacentes. – In: R. J. Fulton (ed.). Le Quaternaire du Canada et du Groenland. Commission glaciologique du Canada, Géolo-gie du Canada, vol. 1: 813-821.

Funder, S. 1989 b. Quaternary geology of the ice-free and adjacents shelves of Greenland. – In: Fulton, R. J. (ed.) Quaternary Geology of Canada and Greenland: 741-792.

Funder, S. & Abrahamsen, N. 1988. Palynology in a polar desert, eastern North Greenland. – Boreas 17: 195-207.

Funder, S. & Hjort, C. 1973: Aspects of the Weichselian chronology in central East Greenland. – Boreas 2: 69-84.

Gams, H. 1931. Die klimatische Begrenzung von Pflanzenarealen und die Verteilung der hydrischen Kontinentalität in den Alpen. – Zeitschrift der Gesellschaft für Erdkunde Berlin 1931: 321-346.

Gams, H. 1932. Die klimatische Begrenzung von Pflanzenarealen und die Verteilung der hydrischen Kontinentalität in den Alpen. II.Teil. – Zeitschrift der Gesellschaft für Erdkunde Berlin 1932: 178-198.

Gartmann, F.R. 1993. Das ökologische Verhalten von Gefässkryptogamen an ihren Ausbreitungsgrenzen in der Arktis und in den Hochalpen. – Dissertation, Universität Zürich: 100 pp.

Geißdörfer, R. 1994. Untersuchungen zur Vertikalverbreitung grönländischer Gebirgsvegetation mit Hilfe der nichtlinearen Regressionsanalyse. – Diplomarbeit, Universität Dortmund: 148 pp.

Gelting, P. 1934. Studies on the vascular plants of East Greenland between Franz Joseph Fjord and Dove Bay (Lat.73°15'-76°20'N). – Meddr Grønland 116 (3): 196 pp.

Gelting, P. 1937. Studies of the food of the East Greenland ptarmigan especially in its relation to vegetation and snow-cover. – Meddr Grønland 116 (3): 196 pp.

Ghisler, M. 1986. Geology and mineral deposits in the National Park in North/ East Greenland. – Forskning/tusaat 3: 43-51.

Gottfried, M., Pauli, H. & Grabherr, G. 1994. Die Alpen im „Treibhaus": Nachweis für das erwärmungsbedingte Höhersteigen der alpinen und nivalen Vegetation. – Jahrbuch des Vereins zum Schutz der Bergwelt 59: 13-27.

Grabherr, G., Gottfried, M. & Pauli, H. 1994. Climate effects on mountain plants. – Nature 369: 448.

Grabherr, G., Gottfried, M., Gruber, A. & Pauli, H. 1995. Patterns and current changes in alpine plant diversity. – In: Chapin, F. S. and Körner, C. (eds.). Arctic and alpine diversity; pattern, causes and ecosystem consequences. Ecol. stud. Springer, Heidelberg: 167-181.

Graham, R. W. & Grimm, E. C. 1990. Effects of global climate change on the patterns of terrestrial biological communities. – Trends Ecol. Evol. 5: 289-292.

Gribbon, P. W. F. 1968. Altitudinal zonation in East Greenland. – Bot. Tidsskrift., 63 (4): 342-357.

Haller, J. 1953. Geologie und Petrographie von West-Andrées Land und Ost-Fraenkls Land (NE-Grönland). – Meddr om Grønland 113 (5): 196 pp.

Haller, J. 1956 a. Geologie der Nunatakker Region von Zentral-Ostgrönland zwischen 72°30′ und 74°10′ N. Br. – Meddr Grønland 154 (1): 172 pp.

Haller, John 1956 b. Die Strukturelemente Ostgrönlands zwischen 74° und 78° N. – Meddr Grønland 154 (2): 27.

Haller, J. 1958. Der „Zentrale Metamorphe Komplex" von NE-Grönland. Teil II. Die geologische Karte der Staunings Alper und des Forsblads Fjordes. – Meddr Grønland 154 (3): 153 pp.

Haller, J. 1970. Tectonic map of East Greenland (1:500 000). An account of tectonism, plutonism, and volcanism in East Greenland. – Meddr Grøn-land 171 (5): 260 pp.

Haller, J. 1983. Geological map of Northeast Greenland 75°-82° N. Lat. 1: 1.000.000. – Meddr Grønland 205 (5): 22 pp.

Halliday, G. 1981. British North-east Greenland Expedition 1980. – University of Lancaster: 20-28.

Halliday, G., Kliim-Nielsen, L. & Smart, I. H. M. 1974. Studies on the flora of the Blosseville Kyst and on the hot springs of Greenland. – Meddr Grønland 199 (2): 49 pp.

Hartz, N. 1895. Fanerogamer e Karkryptogamer fra Nordøstgrønland ca. 75°-70°N.Br. og Angmagssalik, ca. 65°40′N. – Meddr Grønland 18 (5).

Hartz, N. & Kruuse, C. 1911. The vegetation of Northeast Greenland, 69°25′ lat.n. – 75° lat.n. – Meddr Grønland 30 (10): 333-431.

Hegetschweiler, J. 1825. Reisen zwischen Glarus und Graubünden. – Zürich.

Henriksen, N., Perch-Nielsen, K. & Andersen, C. 1980. Geological map of Greenland 1 : 100 000. Sydlige Stauning Alper 71 Ø.2 Nord. The regional geology of a segment of the Caledonian fold belt, and adjacent post-Caledonian rocks, Scoresby Sund region, East Greenland. – The Geological Survey of Greenland, Copenhagen: 46 pp.

Hess, H. E., Landolt, E. & Hirzel R. 1967. Flora der Schweiz und angrenzender Gebiete. – Birkhäuser Verlag, Basel und Stuttgart, Bd. I, 858 pp.

Higgins, A. 1994. Place names of East Greenland 78°-82°N together with a short exploration history. – The Geological Survey of Greenland, Copenhagen: 18 pp.

Holmen, K. 1957. The vascular plants of Peary Land, North Greenland. – Meddr Grønland 124 (9): 149 pp.

Holt, R. D. 1990. The microevolutionary consequences of climate change. – Trends Ecol. Evol. 5: 289-292.

Hovmøller, E. 1947. Climate and weather over the coast-land of Northeast Greenland and the adjacent sea. – Meddr om Grønland 144 (1), Appendix, 1-208.

Hultén, E. 1941-1950. Flora of Alaska and Yukon. I – IX. – Kgl. Fysiogr. Sällsk. Handl. N. F.: 52-61.

Hultén, E. 1956. The *Cerastium alpinum* complex. A case of world-wide introgressive hybridization. – Sv. Bot. Tidsskr. 50 (3): 411-495.

Huntley, B. 1991. How plants respond to climate change: migration rates, individualism and the consequences for plant communities. – Ann. Bot. 67, suppl. 1: 15-22.

Juzepczuk, S. V. 1929. Beitrag zur Gattung *Dryas* L.. – Bull. Jard. b. p. de l'URSS 28: 306-327.

Kaiser, S. 1994. Untersuchung der Artenpräsenz am Beispiel der grönländischen Gebirgsvegetation mithilfe von Diversitätsindices. – Diplomarbeit, Universität Dortmund: 93 pp.

Katz, H. R. 1952. Ein Querschnitt durch die Nunatakkerzone Ostgrönlands (ca.74° N.B.). Ergebnisse einer Reise vom Inlandeis (in Zusammenarbeit mit den Expéditions Polaires Françaises von P.-E. Victor) ostwärts bis in die Fjordregion ausgeführt. – Meddr Grønland 144 (8): 65 pp.

Katz, H. R. 1953. Journey across the nunataks of Central East Greenland. – Arctic 6 (1): 1-14.

Kempter, E. 1961. Die Jungpaläozoischen Sedimente von Süd Scoresby Land (Ostgrönland 71° N) mit besonderer Berücksichtigung der kontinentalen Sedimente. – Meddr Grønland 164 (1): 123 pp.

Kempton, R. A. & Wedderborn, R. W. M. 1978. A comparison of three measures of species diversity. – Biometrics 34: 25-37.

Kempton, R.A. 1979. The structure of species abundance and measurement of diversity. – Biometrics 35: 307-321.

Koch, L. & Haller, J. 1971. Geological map of East Greenland 72°-76° N. Lat. 1:250 000. – Meddr Grønland 183: 13 sheets.

Körner, C. 1994. Impact of atmospheric changes on high mountain vegetation. – In: Beniston, M. (ed.). Mountain environments in changing climates. Routledge, London: 155-166.

Kolderup, A. 1879. Om det organiske Liv på en av Nunatakkerne. – Mddr Grønland 1:(4).

Krebs, C.J. 1989. Ecological Methodology. – Harper & Row, Publishers, New York.

Kruuse, C. 1898. Vegetationen i Egedesminde Skjærgård. – Meddr Grønland 14: 348-399.

Landolt, E. 1964. Unsere Alpenflora (3. Aufl.) – SAC, Zollikon-Zürich: 222 pp.

Mac Arthur, R.H. 1972. Geographical Ecology. – Harper & Row, New York.

Magurran, A. E. 1988. Ecological Diversity and its Measurement. – Croom Helm, London.

Merxmüller, H. 1952. Untersuchungen zur Sippengliederung und Arealbildung in den Alpen. – Verein zum Schutze der Alpenpflanzen und -tiere, München: 105 pp.

Moore, P. D. 1990. Vegetation's place in history. – Nature 347: 410.

Murbeck, S. 1898. Studier över kritiska kärlväxtformer. III. De nord-europeiska formerna af slägtet *Cerastium*. – Bot. Not. Lund.

Nathorst, A. G. 1890. Kritiska anmärkningar om den grönlandska vegetationens historia. – Bih. t. Kungl. Sv. Vet. Akad. Handlingar 16, Afd.III.

Nathorst, A. G. 1891. Fortsatta anmärkninger om den grönlandska vegeta-tionens historia. – Öfvers. of Kungl. Vet. Akad. Förh. 4.

Ohmura, A. & Reeh, N. 1991. New precipitation and accumulation maps for Greenland. – Journal of Glaciology, 37 (125): 140-148.

Ostenfeld, C. H. 1926. The flora of Greenland and its origin. – Dan. Biol. Medd. 6 (3).

Patile, C.P. & Taillie, C. 1982. Diversity as a concept and its measurement. – JASA 77: 548-567.

Peel, J. S. & Sønderholm, S. 1991. Sedimentary basins of North Greenland. – Grønlands Geologiske Undersøgelse. Bulletin 160. København: 164 pp.

Perch-Nielsen, K., Birkenmayer, K., Birkelund, T. & Aellen, M. 1975. – Meddr Grønland 193 (4): 51 pp.

Philipp, M. 1972. The *Stellaria longipes* group in N.W. Greenland. Cytological and morphological investigations. – Bot. Tidsskr. 67 (1-2): 64-75.

Piélou, E. C. 1975: Ecological Diversity. – Wiley, New York .

Porsild, A. E. 1926. Contributions to the flora of West Greenland. – Meddr Grønland 58 (2): 159-196.

Porsild, A. E. 1947. The genus *Dryas* in North America. – Canad. field-naturalist 61 (6).

Porsild, A. E. 1965. The genus *Antennaria* in eastern Arctic and subarctic Canada. – Bot. Tidsskr. 61: 22-55.

Schinz, H. & Keller, R. 1923. Flora der Schweiz. I. Teil: Exkursionsflora (4. Aufl.). – A. Raunstein, Zürich: 792 pp.

Schwarzenbach, F. H. 1951. Ökologische Beiträge zur quartären Florengeschichte Ostgrönlands.- Bericht über das Geobotanische Institut Rübel (Zürich) für das Jahr 1950: 44-46.

Schwarzenbach, F. H. (in Fraenkl 1954). Übersicht über die Botanik und Zoologie von Kronprins Christians Land. – Mddr. Grønland 116 (2): 13-19.

Schwarzenbach, F. H. 1956. Die Beeinflussung der Viviparie bei einer grönländischen Rasse von *Poa alpina* L. durch den jahreszeitlichen Licht- und Temperaturwechsel. – Ber. Schweiz. Bot. Ges. 66: 204-223.

Schwarzenbach, F. H. 1960. Die arktische Steppe in den Trockengebieten Ost- und Nordgrönlands. – Bericht des Geobotanischen Institutes Rübel (Zürich) 38: 41-64.

Schwarzenbach, F. H. 1961. Botanische Beobachtungen in der Nunatakkerzone Ostgrönlands zwischen 74° und 75°N.Br. – Meddr Grønland 163 (5): 172 pp.

Schwarzenbach, F. H. 1996 a. Revegetation of an air strip and dirt roads in Central East Greenland. Arctic 49 (2): 194-199.

Schwarzenbach, F.H., Kaiser, S. & Geissdörfer R. 1996 b. Zur Diversität der natürlichen Gebirgsvegetation Ost- und Nordostgrönlands. – Gaia 5 (3/4): 166-182.

Scoggan, H. J. 1978. The Flora of Canada. – Can. Pubs. Natur. Sci. Botany

Seidenfaden, G. 1933. The vascular plants of South-East Greenland 60°04' to 64°30°N.lat. – Meddr om Grønland 106 (3): 129 pp.

Seidenfaden, G. & Sørensen, T. 1937. The vascular plants of Northeast Greenland from 74°30' to 79°00' N.lat and a summary of all species found in East Greenland. – Meddr Grønland 101 (4): 215 pp.

Simpson, E. H. 1949. Measurement of diversity. – Nature 163: 688.

Söllner, R. 1954. Recherches cytotaxonomiques sur le genre *Cerastium*. – Ber. Schweiz. Bot. Ges. 64:221-354.

Stauber, H. 1940. Stratigraphisch-geologische Untersuchungen in der ostgrönländischen Senkungszone des nördlichen Jamesonlandes. Vorläufiger Bericht. – Meddr Grønland 114 (7): 34 pp.

Stugren, B. 1986: Grundlagen der Allgemeinen Ökologie, 4. Aufl. – Gustav Fischer Verlag, Stuttgart: 356 pp.

Sørensen, T. 1933. The vascular plants of Northeast Greenland from 71°00' to 73°30'N lat. – Meddr Grønland 101 (3): 177 pp.

Sørensen, T. 1941. Temperature relations and phenology of the Northeast Greenland flowering plants. – Meddr Grønland 125 (9): 305 pp.

Tallis, J. H. 1991. Plant community history. – Chapman & Hall, London. 398 pp.

Theurillat, J.-P. 1995. Climate change and the alpine flora: some perspectives. – In: Guisan, A., Holten J. I., Spichier, R. & Tessier, L. (eds.). Potential ecological impacts of climate change in the Alps and Fennoskandian mountains. – Editions Convervatoire et Jardin Botanique Genève: 121-127.

Urfer, W. & Schwarzenbach, F. H. 1995. Analysis of arctic mountain vegetation: diversity and vertical distribution. – Coenosis 10, (2/3): 133-140.

Vaage, J. 1932. Vascular Plants from Eirik Raude's Land (East Greenland 71°30'-75°40'lat. N). – Skrifter om Svalbard og Ishavet 49: 87 pp.

van Steenis, C. G. G. 1961. An attempt towards an explanation of the effect of mountain mass elevation. – Proc. Lon. Ned. Akad. Wetensch. C 64: 435-442.

Warming, E. 1888. Om Grønlands vegetation. – Meddr Grønland 12: 245 pp.

Washington, H. G. 1984. Diversity, biotic and similarity indices: a review with special relevance to aquatic ecosystems. – Water Research 18: 653-694.

Wilson, E. O. & Peter, F. M. (eds.) 1988: Biodiversity. – National Academy Press, Washington DC.

Witzig, E. 1954. Stratigraphische und tektonische Beobachtungen in der Mestersvig-Region (Scoresby Land, Nordostgrönland). – Meddr Grønland 72, 2. Avdeling (5): 26 pp.

Yurtsev, B. A. 1994: Floristic division of the Arctic.- Journal of Vegetation. Science 5: 765-776

15. Topographical names

Adolf Hoel Gletscher
Agardh Bjerg
Alpefjord
Amdrup Land
Andrée Land
Angmagssalik
Bartholin Land
Benjamin Dal
Bernhard Studer Land
Bersærkerbræ
Biskop Alf Dal
Biskop Alf Gletscher
Blosseville Kyst
Blåbær Gletscher
Blåbærdal
C. H. Ostenfeld Nunatak
Caledonia Ø
Campanuladal
Cecilia Nunatak
Centrum Sø
Clavering Ø
Daneborg
Danmark Fjord
Danmarkshavn
Dansketinden
Djimphna Sund
Drømmebjerg
Døde Slette
Eleonore Bugt
Elisabeth Bjerg
Ella Ø
Endeløs Gletscher
Eremitdal
Femdalen
Flyverbjerg
Flødegletscher
Forsblad Fjord
Frigg Fjord
Frihedstinde
Fyn Sø
Fynselv
Gauligletscher
Gauss Halvø
Geologfjord
Gerard de Geer Gletscher
Germania Land
Giesecke Bjerge
Grejsdal
Grønnemark
Gullygletscher
Gurreholm Bjerge
Gåseland
Hekla Sund
Henrik Kröyer Holme
Hjørnedal
Hobbs Land
Hochstetter Forland
Holbæk Bugt
Hold with Hope
Holm Land
Hudson Land
Hvidevæggen
Højedal
Ingolf Fjord
Ismarken
Jameson Land
Jordanhill
Junctiondal
Kangerlussuaq
Kap Dalton
Kap Farvel
Kap Hedlund
Kap Peterséns
Kap Weber
Keglebjerg
Kejser Franz Joseph Fjord
Kempe Fjord
Kong Oscar Fjord
Kronprins Christians Land
Krumme Langsø
Langelv
Linné Gletscher
Lomsø
Luciadal
Lyell Land
Længselsbjerg
Marianne Nunatak
Marmorvigen
Mesters Vig
Moraenedal
Mt. Eremit
Mt. Forel
Myggbukta
Mørkefjord
Månesletten
Månevig
Nathorst Land
Nordfjord
Norsketinden
Nunatak Gletscher
Nysne Gletscher
Nøkkefossen
Nørlund Alper
Ole Rømer Land
Ostenfeld Land
Peary Land
Pingo Dal
Pingo Pass
Polhem Dal
Posten
Primulabugt
Promenadedal
Randspids
Renbugt
Rhoess Fjord
Rivieradal
Romer Sø
Rondenæs
Schaffhauserdalen
Schuchert Dal
Scoresby Sund
Scotstounhill
Sedgwick Gletscher
Sedimentkløft
Sefstrøm Gletscher
Segelsällskapet Fjord
Shannon Ø
Silberhorn
Skeldal
Skjoldungebræ
Skjoldungen
Skærfjord
Solvig
Spærregletscher
Station Nord
Stauning Alper
Stenø Land
Strindberg Land
Syltoppene
Sæfaxa Sø
Sæfaxi Dal
Sæfaxi Elv
Sødal
Teufelsschloss
Traill Ø
Tærskeldal
Tågefjeldene
Ubaliq
Ulvebjerg
Vandredalen
Vibeke Nunatak
Vibeke Sø
Vikingebræ
Waltershausen Gletscher
Waltershausen Nunatak
Werner Bjerge
Ymer Ø
Zackenberg